Beautiful China 美丽中国 1

Residential
Landscape 住宅景观

佳图文化 编

华南理工大学出版社
·广州·

图书在版编目（CIP）数据

美丽中国 1：住宅景观 / 佳图文化编 . —广州：华南理工大学出版社，2013.10
ISBN 978-7-5623-3995-3

Ⅰ . ①美⋯ Ⅱ . ①佳⋯ Ⅲ . ①住宅 - 景观设计 - 作品集 - 中国 - 现代 Ⅳ . ① TU241

中国版本图书馆 CIP 数据核字（2013）第 165402 号

美丽中国 1: 住宅景观
佳图文化 编

出 版 人：韩中伟
出版发行：华南理工大学出版社
　　　　　（广州五山华南理工大学 17 号楼，邮编 510640）
　　　　　http://www.scutpress.com.cn　E-mail: scutc13@scut.edu.cn
　　　　　营销部电话：020-87113487　87111048（传真）
策划编辑：赖淑华
责任编辑：杨爱民　庄 彦　赖淑华
印 刷 者：广州市中天彩色印刷有限公司
开　　本：1016mm×1370mm　1/16　印张 :16.5
成品尺寸：245mm×325mm
版　　次：2013 年 10 月第 1 版　2013 年 10 月第 1 次印刷
定　　价：296.00 元

　　　　版权所有　盗版必究　　　印装差错　负责调换

Preface | 前言

Landscape is traces of nature and human activity on the earth and natural landscape is a miracle created by nature while human landscape is drawn down by designer souls. Excellent designers will not debase the nature wonder in their pursuit of soul in pen; instead they will utilize the wonder to create various styles of human landscape.

The shared goal of landscape in American style, British style or Chinese style is to create a graceful dwelling environment. It refers to aesthetics in visual enjoyment and the harmony between human and human, human and environment. For this reason, it requires more than working on external landscaping; designers shall extend their study further to the inner character and definition of landscape, i.e. presenting landscape ecological character, environmental friendly concept etc. A landscape external decides how many people are attracted to come while its inner character decides how many people will really choose here.

A new concept of Beautiful China is firstly presented in the 18th CPC National Congress. It covers two connotations: the beauty of national nature environment and the beauty of national spirit and social development. Landscape planning and design of the time has departed from the single pursuit of landscape external, focuses more on the inner character of ecology, environment-friendly etc. On the basis of ensuring the landscape external appearance, designers pay attention to the selecting materials of low energy consumption, high efficiency and ultra environment-friendly, fully following the concept of "construction of ecological civilization in a prominent position" proposed in the 18th CPC National Congress. The requirements of landscape planning and design of the time tally right with the spiritual connotation of Beautiful China. The book has selected projects correspond to the double standards of Beautiful China, offering a feast of visual beauty as well as an ecological feast of spirit to single person, society and nation.

The book has professionally analyzed the selected project in multiple perspectives. In the content scheduling, the key points, highlights, design concepts etc are analyzed with a great quantity of professional technical drawings, detailed and informative. We expect to help designers and landscape practitioners create more beautiful landscapes in China, achieve Chinese dream of jointly constructing beautiful China and enduring development.

景观，是大自然和人类活动在大地上的烙印。自然景观是大自然创造出来的神奇，人文景观是设计师笔下的灵魂。好的设计师在追求笔下灵魂的时候不仅不会用手中的笔去掉大自然的神奇，而且还会充分利用大自然的神奇创造出各种风格的人文景观。

但不管是美式风格、英伦风格还是中式风格的景观，共同的目标都是创造一个美丽的栖居环境。这里所谓的"美丽"，一是指人视觉上美的享受；二是指人与人之间、人与环境之间的和谐之美。为了实现这个目标，纯粹地从景观外在美化上下功夫是远远不够的，设计师的笔触应该更深入地伸向景观内在属性的定义上，比如赋予景观生态特质、环保理念等。景观的外在决定了能吸引到多少人，而景观的内在属性却决定了真正有多少人选择这里。

中共十八大首次提出"美丽中国"的新概念。"美丽中国"意涵着两个价值维度：一是国家自然环境之美；二是国家精神及其社会发展之美。当代景观规划和设计已经脱离了纯粹的景观外在化追求，而更加注重生态、环保等景观的内在属性。设计师在保证景观外在美观的基础上，注重选用低能耗、高效能、超环保的材料，全面贯彻中共十八大"把生态文明建设放在突出的位置"的精神理念。当代景观规划设计的要求正契合了"美丽中国"的精神内涵。本书在案例选择上执行的就是"美丽中国"的双重标准，既给人带来一次美的视觉盛宴，又让个人、社会和国家在精神上享一场生态盛飨。

本书从多角度详细且专业地分析了"美丽"的景观案例。内容编排上，分别从景观案例的关键点、亮点、设计思想等方面入手，配合大量的专业技术图纸，资料丰富而详实。我们的努力是为了让设计师及景观从业者在中国的蓝图上创作出更多更美的景观，在景观设计上共同实现建设美丽中国、实现中华民族永续发展的中国梦。

CONTENTS 目录

Modern Style 现代风格

The Upriver of Xiangjiang, Changsha ········ 002
长沙湘江一号

Country Garden Holiday Islands Flowers Huadu 21 Third Street ········ 012
花都碧桂园假日半岛鸟语花香三街 21 号

Jiuzhou New World City, Changzhou ········ 018
常州九洲新世界

China Overseas Yuelangyuan, Shenzhen ········ 026
深圳中海月朗苑

Vanke Town Landscape in Zhongshan ········ 034
中山万科城市风景

Hangzhou Gemdale Zizai Town ········ 040
杭州金地自在城

Nantong One House Mansion ········ 048
南通万濠华府

Gemdale Green County, Shanghai ········ 056
上海金地格林郡

Hangzhou Lianhe Geli ········ 060
杭州联合格里

Changsha Vanke Golden Mansion Phase One and Two ········ 066
长沙万科金域华府一、二期

ZOBON City Villa, Zhuhai ········ 074
珠海中邦城市美墅

Neoclassical Style 新古典主义风格

Jinke·Oriental Palace ········ 084
金科·东方王府

Xingjin · Mansion of Shang ········ 090
兴进·上郡

Ocean City, Zhongshan ········ 098
中山远洋城

Jiangsu Changzhou Future Golden County ········ 104
江苏常州新城金郡

Changzhou Xiangyi Zijun New Town ········ 112
江苏常州新城香溢紫郡

CIMC Unique Wenchang Landscape Design ···················· 122
扬州·中集紫金文昌景观设计

Xi'an Jinhui Swan Bay ···················· 128
西安金辉天鹅湾

Shanghai Palaearctic Sheshan Villa ···················· 134
上海古北佘山别墅

Zigong BRC. Gongshan NO.1 (Phase II) ···················· 146
自贡蓝光·贡山壹号（二期）

New Chinese Style 新中式风格

Huidong International Garden, Guangzhou ···················· 154
广州汇东国际花园

Royal Dragon Bay ···················· 160
御龙湾

Landscape Design of Beijing Poly Oriental Mansion ···················· 168
北京保利东郡项目景观设计

Zone P1+M3 of OCT East, Shenzhen ···················· 174
深圳东部华侨城 P1+M3 区

Foshan Hefeng Yingyuan Garden ···················· 182
佛山和丰颖苑

Zhangtai Wisdom City, Guilin ···················· 192
桂林彰泰睿城

Yili Renhe - Ningyuan County ···················· 198
伊犁仁和·宁远郡

Taigu City Garden Residence, Shenzhen ···················· 206
深圳太古城花园

Oasis Lakeside Villa, Shanghai ···················· 216
上海绿洲江南园

Guangzhou Dayi Villa ···················· 224
广州大一山庄

Yangjiang Danmo Cabin ···················· 228
阳江淡墨幽居

Vanke Rain Garden, Nanchang ···················· 236
南昌万科润园

Southern Orchid Garden, Xi'an ···················· 246
西安兰亭坊南区

Modern Style
现代风格

Elegance
简约时尚

Functionalism
功能主义

Modernism
现代格调

MODERN STYLE
现代风格

KEY WORDS 关键词

FALLINGWATER VILLA
流水别墅

NATURAL LIFE
自然生态

LANDSCAPE NODE
景观节点

Location: Changsha, Hunan
Landscape Design: L&A Design Group

项目地点：湖南省长沙市
景观设计：奥雅设计集团

The Upriver of Xiangjiang, Changsha
长沙湘江一号

FEATURES 项目亮点

The design emphasizes the co-exitance and harmony between landscape, architecture and nature, well showing people's respect to nature.

整体设计强调景观与建筑以及自然的相融共生，充分表现了人对于自然的尊重与崇尚。

Site Plan 总平面图

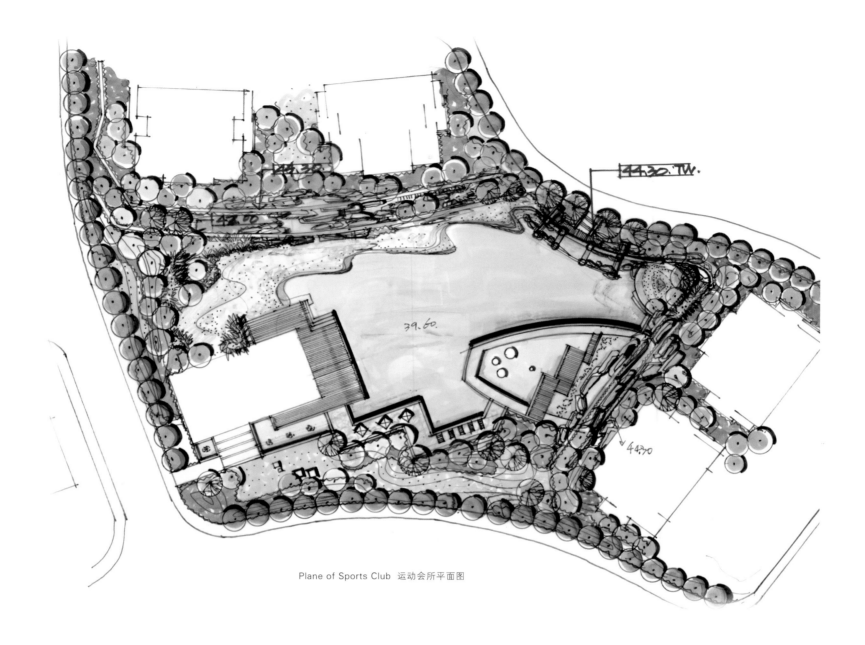

Plane of Sports Club 运动会所平面图

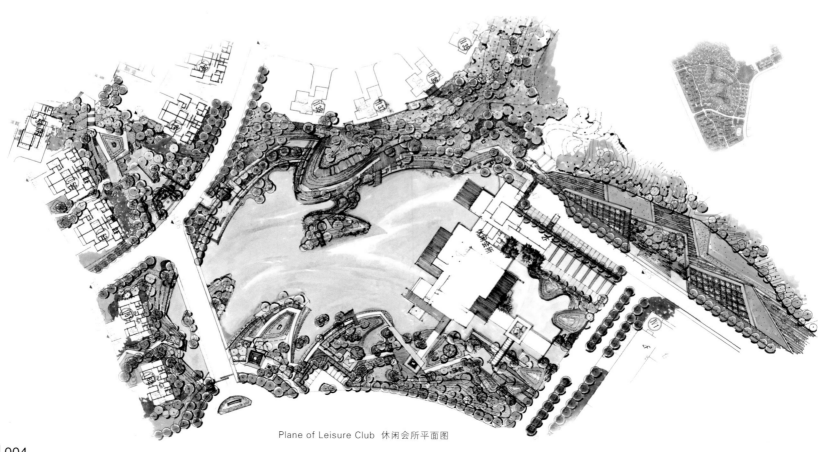

Plane of Leisure Club 休闲会所平面图

Overview 项目概况

The site is sitting on the east bank of Xiangjiang River with Binjiang Avenue on the west, Furong Avenue on the east, Eyangshan Park on the north and the North 2nd Ring Road on the south. It features a 1,000 m long shore line on the west.

The project is in a hilly area which is higher in the north and lower in the south. The site looks like a polygonal "drunkard's wine bottle" with the bottom faces to Xiangjiang River and the "bottle neck" faces to Furong North Road. There is plentiful splendid eco vegetation within the site. The project faces to Xiangjiang River on the west with an open view. It also features rich vegetation resource, water resource, and profound cultural context.

The architectural design has drawn the essence from Frank Lloyd Wright's design idea "Living with nature harmoniously", to use a great number of horizontal lines and local materials, and keep harmonious with nature. Wright's Prairie Houses and Fallingwater just show people's respect, love and curiosity to nature.

项目位于湖南长沙湘江东岸，西面紧邻滨江大道，东面与芙蓉大道相接，北靠鹅羊山公园，南面与市北二环线紧邻，地块西侧有 1 km 临江面。

项目总体为北高南低的丘陵山地，呈多边形的"酒鬼酒瓶"状，朝向湘江一侧是"瓶底"，朝向芙蓉北路一侧是"瓶嘴"，其内有多处美轮美奂的原生态植物群落。项目西临湘江，视野开阔，原生植被及水资源丰富，地缘（鹅羊山）历史文脉深厚。

本项目的建筑设计吸取了赖特对建筑设计的精髓：从大量横向线条的使用、乡土材料的运用、与自然相结合等设计可看出，设计师把建筑看作对自然界的敏感回应，赖特设计的草原式住宅和流水别墅，表现了人特有的对自然的尊重和崇尚，反映了人们对自然的热爱和好奇。

Design Concept 设计理念

The landscape design has paid attention to "natural landscape", "life" and "organic architecture". It aims to create an ideal living environment which is caring, ecological and healthy with cultural ambiance.

整个项目的景观设计强调"自然山水"、"生命"和"有机建筑"。项目景观设计将全力创造一个人性化、自然生态化、健康化以及人文化的理想居住环境。

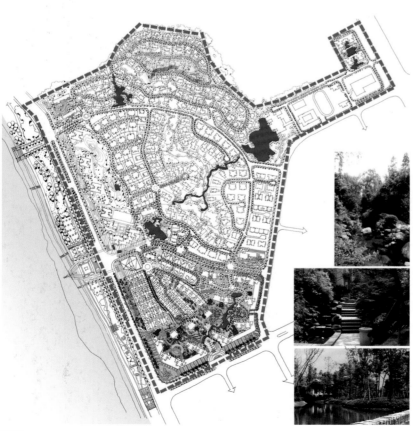
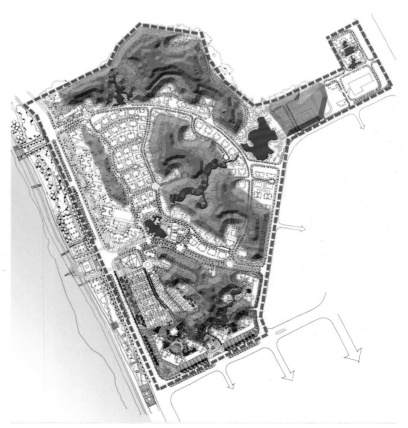

▶ Landscape Node 景观节点

Natural Landscape: the design has taken advantages of the surrounding environment to create a high-end neighborhood with "water" and "green" by adopting natural, organic and green materials.

Life: it emphasizes modest scale and comfortable design. Both the public space and private space are designed with aesthetics, which enables the residents to communicate with each other in the beautiful environment. And the hills in the center have provided the residents an ideal place for climbing and outdoor activities.

Organic Architecture: the father of organic architecture Frank Lloyd Wright had ever said that it is much better to "build in the mountain" than "build on the mountain". The landscape design actually aims to create this kind of environment to make the architecture "grow" on the site.

自然：该项目配合周边的自然山势与环境，在"自然天成"的主题下，选用自然的、有机的材质，通过绿化的衔接，成功打造出一个与自然相结合的、"水""绿"交融的高尚住宅区。

生命：强调人性尺度和宜人的设计，在人的活动范围内，无论是公共空间还是私人空间都有美学渗透，有利于住户在社区的优美环境中拉近彼此距离，进行充分的交流。中心山体公园为住户提供了登山健身、户外活动的理想场所。

有机建筑：有机建筑之父赖特先生说过，把一座建筑与场地的关系表达为"建于山中"要比"建在山上"好得多。景观的设计实质上就是在营造这种场地，使得建筑看起来是从场地上生长出来的，是场地独一无二的结果。

MODERN STYLE
现代风格

Country Garden Holiday Islands Flowers Huadu 21 Third Street
花都碧桂园假日半岛鸟语花香三街21号

KEY WORDS 关键词

ABSTRACT ART
抽象艺术

COURTYARD SPACE
庭院空间

LANDSCAPE NODE
景观节点

Location: Huadu District, Guangzhou, Guangdong
Landscape Design: Guangzhou Yuanmei Design
Total Land Area: 6,670,000 m²

项目地点：广东省广州市花都区
景观设计：广州市圆美环境艺术设计有限公司
总占地面积：6 670 000 m²

FEATURES 项目亮点

Simple and elegant design creates a comfortable, green, healthy living environment with the courtyard as the core element.

设计以简约的形式展现了一个舒适、绿色、健康的居住环境，庭院空间的设计是整个项目的核心。

Design Concept 设计理念

This is a modern-style courtyard which emphasis on a simple beauty pursing free, unrestrained and generous atmosphere. In terms of functional design, it creates a comfortable, green, healthy living environment and features both rich cultural atmosphere and local artistic characteristics.

　　此设计是现代式庭院，主要讲究的是一种简约之美，追求自由、奔放和大气，大则幽深静谧，小则精致秀巧。在功能设计等方面营造出一个舒适、绿色、健康、富有文化气息和地方艺术特色的居住环境。

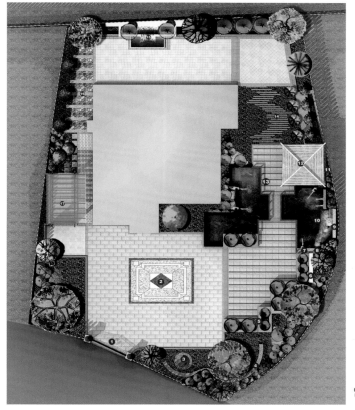

❶ 别墅入口	❻ 文化石景墙	⓫ 亭背景景墙	⓰ 过渡空间
❷ 拼花铺装	❼ 吐水小品	⓬ 现状凉亭	⓱ 上抬花架
❸ 吐水小景	❽ 散置鹅卵石	⓭ 眺水平台	
❹ 孤行矮景墙	❾ 跌水景墙	⓮ 条形汀步	Master Landscape Plan
❺ 错级花池	❿ 深水鱼池	⓯ 多级跌水景	景观总平面图

▶ Courtyard Landscape 庭院景观

Courtyard in front of the residence is the transition zone to the outside world, but also an extension of the family space. A full-featured garden with friendly environment not only allows you to feel the natural world without stepping out, but to dine, chat with family or friends inside.

Regarding to landscape elements, simple rectangular, circular and trapezoidal looked beautiful with no lack of practicality. Abstract sculpture and art flowerpot are the main decorative elements in the garden, besides, there are natural elements such as stones, pebbles, wood, etc. Tall and narrow lines are formed by tall trees configuring with low plants to achieve visual balance. In addition, bamboo with other plants decorates partial areas.

　　住宅前的庭院，是家庭向外界过渡的地带，也是家庭空间的延伸。小小庭院好处多，一个环境好、功能齐全的庭院，不仅可以让人们足不出户就可以轻松回归自然，也可以与家人或朋友坐在其中用餐、聊天，放松心情。

　　景观元素主要是采用简单的长方形、圆形和梯形等，既美观大方，又不乏实用性。庭院中的构筑物形式简约，采用抽象雕塑品、艺术花钵等为庭院的主要装饰元素，同时还运用一些天然的元素，例如石块、鹅卵石、木板等。庭院绿化用垂直高大的乔木形成高大、狭长的线条，配置低矮的植物来达到视觉上的平衡。此外还使用竹子等植物装饰局部景观。

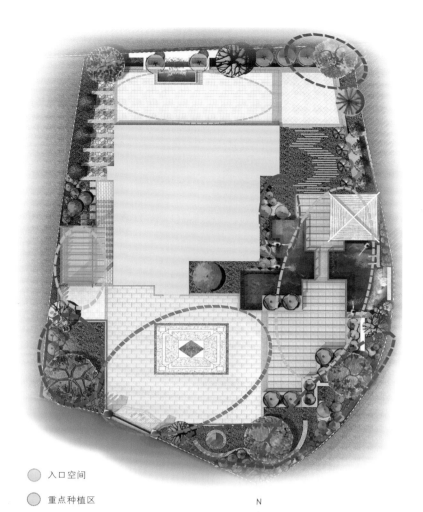

- 入口空间
- 重点种植区
- 凝思空间

Functional Analysis Drawing
功能分析图

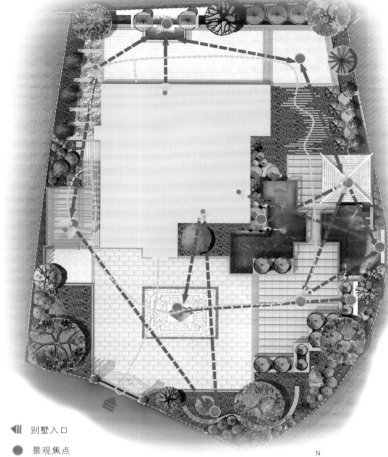

- 别墅入口
- 景观焦点
- 对景面
- 园内步径
- 景观视角

Landscape View Analysis
景观视线分析图

015

MODERN STYLE
现代风格

KEY WORDS 关键词

ECOLOGICAL SYSTEM
生态系统

GROUP VERDURIZATION
组团绿化

LANDSCAPE NODE
景观节点

Location: Changzhou, Jiangsu
Developer: Jiangsu Jiuzhou Investment Group
Landscape Design: EDSK
Designer: Sun Lihui, He Xin, Zheng Huiguo, Wang Hangdong
Land Area: 280,000 m²

项目地点：江苏省常州市
开 发 商：江苏九洲投资集团
景观设计：美国 EDSK 易顿国际设计集团有限公司
主要设计人员：孙立辉 何鑫 郑辉国 王航东
用地面积：280 000 m²

Jiuzhou New World City, Changzhou
常州九洲新世界

FEATURES 项目亮点

The project adopts the gardens' layout and the modern elemental style, framing the beautiful life of "enjoying the nature in the center of the city".

项目采用山水园林的布局与当代高尚元素相融合的设计风格，构架"居城市之心，享山水之情"的美景生活。

Overview 项目概况

Located in Tianning District of Changzhou with planned land area of 175,000 m², the project is a to-be-developed site with the largest scale, the most composite functions and the deepest influence. It is also the largest reconstruction project of Jiuzhou Group. As the reconstruction from old town, the west commercial area has mature commercial atmosphere. Jiuzhou Group aims at creating an international garden living community with humanity ecology.

九洲新世界位于常州市天宁区块，规划土地面积约175 000 m²。是常州中心城市核心区域规模最大、功能复合程度最高、影响最大的待开发用地，对于九洲集团来说也是目前最大规模的拆迁改造项目。基地西侧大型商业区作为旧城改造项目区域商业氛围成熟，九洲集团致力于将其打造成一个国际人文高尚生态的花园生活社区。

Site Plan 总平面图

▶ Design Concept 设计理念

According to the land condition of Changzhou, the project adopts the gardens' layout and the modern elemental style, building the high-quality living environment that respects and conforms to the nature and emphasizing the harmony and interaction between human and nature with the smooth group. The modern urban life is displayed, framing the beautiful life of "enjoying the nature in the center of the city". The core concept of water design aims at reaching a perfect and humane relation between human and nature, human and human, human and construction, construction and nature.

鉴于常州市及其基地现状，项目设计采用山水园林的布局与当代高尚元素相融合的设计风格，营造尊重自然、顺应自然的高品质居住环境，并以非常流畅的组团形式来强调人与自然、人与人之间的和谐与交流，共同演绎现代都市生活的经典，构架"居城市之心，享山水之情"的美景生活。项目以水为设计的核心理念就是使人与自然、人与人、人与建筑、建筑与自然达成最理想、最人情化的互动关系。

▶ Landscape Node 景观节点

The project creates the most natural beauty of spirituality by putting the Chinese classic landscape layout into the core of the flourishing city. In order to guarantee the effect of the landscape design and based on the location, the designers use the natural and ecological waterscape in which the fishes are swimming and the plants are growing, building a balanced waterside ecological system.

The design allows people to see the scenery by pushing the window and the water by looking down. The landscape design is considered comprehensively combined with the residential area. The central group verdurization becomes the connection for the roads and the waters. The riffles and the watercourses are set layers by layers in coordination with the landscape nodes. Therefore, the excellent ecological environment consisting landscape and objects is made.

The plant landscape design emphasizes on natural style with changes of the seasons, all kinds of plants, forming a scenic garden where people can both visit and rest. The public area is designed mainly with landscapes, emphasizing the safety facilities for the aged and children and the protection for the ground-floor residents, which presents the design concept of people-orientation.

The trees along both sides of the lanes are designed according to the change of the seasons and the demand for lighting. Deciduous trees are planted along the east-west-oriented roads to make sure sunlight will be available in the garden in winter; while evergreen trees are set along the south-west-oriented roads. All kinds of local trees are used to enrich the landscape along the roads, soften the constructions and showing landscapes in all seasons. The surrounding verdurization can block the strangers, eliminate the exhaust gas, decline the noise and build the landscapes. There are lots of water plants in the water system landscape and various flower pots in the garden.

九洲新世界创造了一片最自然的灵性之美，以中国古典山水为布局的山水绿洲置于繁华城市的核心。为了保证水景设计的效果，设计师考虑到基地的纬度偏北，采用自然生态的绿色水景设计，池中放观赏鱼，局部草坡入水处理，水岸边种植水生植物，营造出一个平衡的水岸生态系统。

项目运用推窗即见景、低首即望水的设计理念，景观设计结合住宅区通盘考虑，中心组团绿化成为连接各种形态路面和水面的枢纽。各交汇处浅滩、河道配合设置的局部景点，层层递进，从而达到景中有物、物中有景、环抱围合的优质生态环境。

九洲新世界的植物景观设计以突出自然风情为主，强调季节的变化，种植层次丰富、品种多样的花草树木，形成可观、可游、可憩的风景园林。公共组团区的设计以山水为主，强调了老人与儿童活动的安全设置以及对首层住户私密性的保护，体现了以人为本的设计理念。

小区车道两侧林荫设计考虑季节的变化与采光的要求，东西向道路用落叶乔木，以保证花园冬季的阳光；南北向道路用常绿乔木。运用当地多种常绿、落叶树混植，以丰富道路景观，柔化建筑，使道路四季有景。小区外围绿化具有阻隔外人、消除废气、降低噪音、营造区内景色的多重作用。园区的水系景观中配置丰富的水生植物，另外园区布置了形式多样的盆栽花钵。

MODERN STYLE
现代风格

KEY WORDS 关键词

MULTIPLE SPACE
多元空间

INNOVATIVE COMBINATION
创意组合

LANDSCAPE NODE
景观节点

Location: Shenzhen, Guangdong
Developer: China Overseas Land Co., Ltd.
Landscape Design: Shenzhen Art-spring Landscape Design Co., Ltd.
Landscape Area: 48,000 m²

项目地点：广东省深圳市
开 发 商：中海地产股份有限公司
景观设计：深圳市阿特森泛华环境艺术设计有限公司
景观面积：48 000 m²

China Overseas Yuelangyuan, Shenzhen
深圳中海月朗苑

FEATURES 项目亮点

The project has followed new modernism art style, which is natural, simple and elegant, to create a modern and elegant atmosphere and show the strong flavor of the ocean.

本案设计承袭自然简约的现代艺术风格，将外部景观与建筑内部结构进行有机结合，打造了一个具有浓郁海洋气息的现代景观空间。

Overview 项目概况

Located in Bantian Village, Buji Town, Longgang District, Shenzhen, the project is 4,000 m away from Meilin Gate. The site is in rectangle shape with a 5.5 m elevation difference between southeast corner and northwest corner. Vanke The fifth Village is on the east side, Jigong Mountain is on the south side, and Yayuan Road is on the north side. There are multi-storey buildings, medium-rise buildings and high-rise buildings as well as a 20,000 m² central garden to create a comfortable, natural and leisurely living atmosphere. All the flats are designed with big width and small depth to enjoy beautiful landscape. Each of them features large home garden or landscape terrace and huge French window which provide open view. The living spaces are arranged flexibly to create new-type and healthy residences.

该项目位于深圳市龙岗区布吉镇坂田村，距梅林关口约 4 km，用地基本为矩形，东南角与西北角高差约为 5.5 m。用地东侧紧邻万科第五园项目，南侧为鸡公山，北侧为雅园路。住宅楼型为多层、小高层、高层相结合，拥有近 20 000 m² 的中心花园，展现一种宽适、自然、休闲的生活氛围。所有户型均为大宽面、浅进深设计，户户有景、户户赠送超大挑高入户花园或观景露台及超大落地凸窗，更显豁然通透，生活空间自由创意组合，完美呈现新型健康时尚居所。

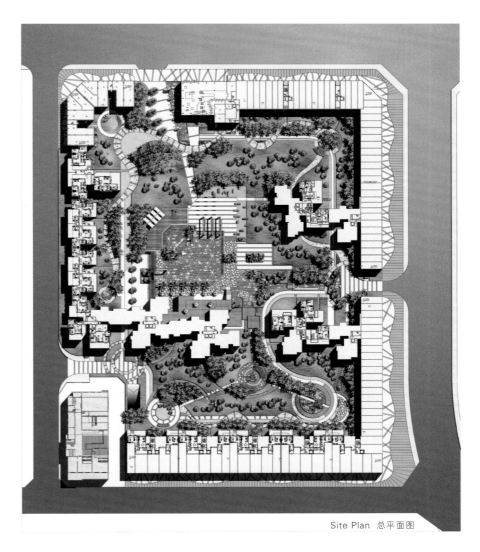

Site Plan 总平面图

> Design Goal 设计目标

As the site has a plentiful of levels and limited factors, the designer seeks for the best design inspiration in the design process according to these problems, and creates a lively environment with the combination of human, ecosystem and artistic landscape. The designer has adopted many types of space combinations to create the atmosphere of "home" not only for the owners but also for the visitors.

In the master plan, as the site has a plentiful of levels and limited factors, the design should be natural, simple, delicate and full of connotation without multifarious elements. This is also the soul of the design.

项目地块有着非常丰富的标高，但相对受限因素较多，鉴于这些背景问题，设计师在专业景观设计的过程中寻求最佳设计灵感，赋予该地块以人、生态、景观艺术相互融合的生态环境。设计师采用创建多种空间组合的方式，不仅为该设计的"最终享有者"营造出一种浓厚的"家"的氛围，同时也为到此参观的游人们提供了别致的家的情意。

在总平面图中，由于地块的受限，在设计过程中不允许过于繁杂的设计元素，取而代之的是自然简洁、精致且富有内涵的创作风格，这也是该设计的灵魂理念。

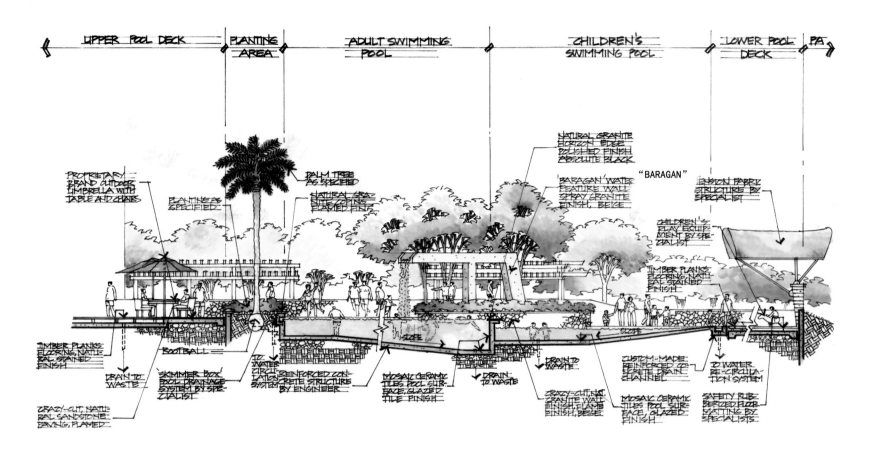

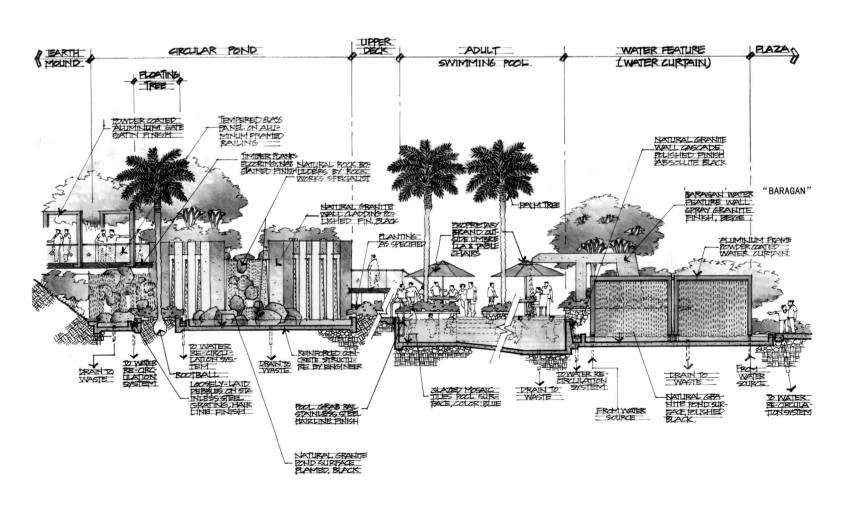

▶ Landscape Feature 景观特色

Combining exterior landscape elements with the interior spaces of the buildings is one of the important aims. For example, the ground floors of the buildings are elevated. This design idea is penetrated into the practical design and developed to natural and pure creation which makes the project close to the heart of the owners.

With green tropical garden as the background, streams are flowing on the huge boulders, and the fresh fruits hang on the dense trees to create spiritual landscape. Beautiful flowering shrubs, and well-trimmed lawns are set along the path to create a natural and cozy living space.

The project has followed new modernism art style which is natural, simple and elegant. Elegant, fresh and bright colors as well as pleasant natural landscapes are adopted to create a modern and elegant atmosphere and show the strong flavor of the ocean. The "ventilating" buildings have provided the residents with maximum comfort and create a pleasant and high-quality atmosphere for leisurely life.

　　融合外部景观元素与建筑物内部结构亦是项目的重要目标之一，例如较大面积的建筑底部架空层的处理，这种灵魂理念深入到现实设计中，并逐渐演变成了自然纯真感觉的创作过程，使项目进一步贴近广大住户的心。

　　在青葱的热带园林的背景下，潺潺细泉淙淙滑过大漂石，浓密的树木垂挂着晶莹剔透的四季水果，让景观变得更加富有灵气。花团锦簇的灌木丛、修葺整齐平整的草坪散落在漫步道的两侧，展现出自然惬意的生活空间。

　　中海月朗苑承袭自然简约的新现代艺术风格，大方、清丽、明亮的色彩，融入时尚精致的高雅意蕴，舒适宜人的自然景观等，展现出具有浓郁海洋气息的蓬勃朝气，最大限度地体贴居者的生活舒适度的"透气"建筑，营造出一种宽适的休闲生活及品质社区氛围。

MODERN STYLE
现代风格

KEY WORDS 关键词

GROUP LANDSCAPE
组团景观

MULTI LEVEL
多重层次

LANDSCAPE NODE
景观节点

Location: Zhongshan, Guangdong Province
Developer: Zhongshan Vanke Real Estate Co., Ltd.
Landscape Design: Shenzhen Yalantu Landscape Engineering Design Co., Ltd.

项目地点：广东省中山市
开 发 商：中山市万科房地产有限公司
景观设计：深圳雅蓝图景观工程设计有限公司

Vanke Town Landscape in Zhongshan
中山万科城市风景

FEATURES 项目亮点

The project pays attention to restore beautiful natural landscape, creating a multi-level landscape system of natural, ecological and harmonious style.

本案注重对自然美景的重塑，打造具有丰富层次感的社区景物，设计风格追求自然、生态与和谐。

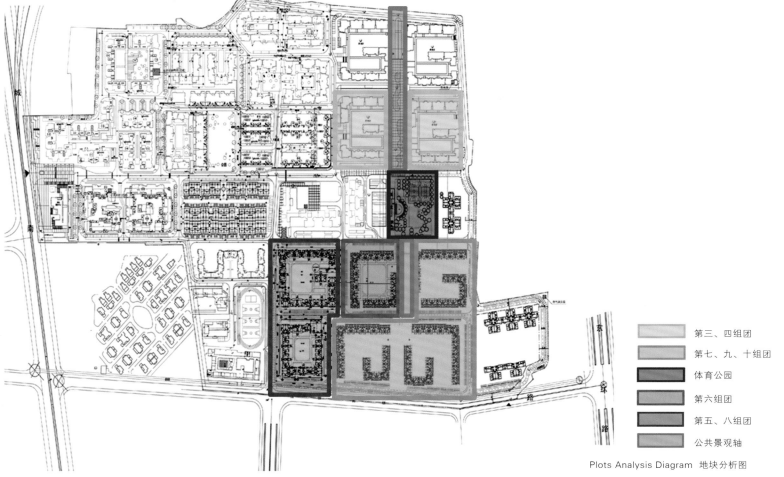

	第三、四组团
	第七、九、十组团
	体育公园
	第六组团
	第五、八组团
	公共景观轴

Plots Analysis Diagram 地块分析图

▶ Overview 项目概况

The project is located in the central area of South District of Zhongshan City in Guangdong Province, next to Shiqi District and East District, with complete public service facilities and convenient transportation. This part of design includes the seventh, ninth and tenth cluster, covering a total area of 100,000 square meters. The style of these three clusters is in consistent with previous finished clusters and with distinct features as well.

项目位于广东省中山市南区中心区，与石岐区、东区相邻，周边公共服务设施配套齐全，交通极其便利。本次设计为整个项目的第七、九、十组团，面积约 10 万 m²，景观设计风格继续延续前面已做设计的组团风格，同时又具有自己的鲜明特色。

▶ Design Concept 设计理念

In design concept, architects try to remould the natural landscape and reproduce the changing scenery of nature to provide a visually aesthetic feeling and to enable the visitors to experience the beauty physically. The design style pursues natural atmosphere and harmony. No matter which angle visitors stand in, they could always discover unintentionally the ingenuity of architects and feel the promotion of them to the new urban-rural lifestyle.

在设计理念上，设计师试图通过对自然美景的重塑，再现自然界变化万千的景物，以达到不仅在视觉上给人以美感，而且在空间上提供令人身临其境的目的。设计风格追求自然、和谐，无论在哪个角度，人们都可能在不经意间发现设计师的独具匠心，从而感受到设计师对新都市田园生活方式的倡导。

Landscape Detail 景观细部

In design details, the project pays attention to the integrity and emphasizes on greening and strengthens the visual effect of plants and spatial atmosphere. Meanwhile, it also lays stress on the participation of residents and practicability to reduce the cost on condition of guaranteeing the landscape quality of seventh, ninth and tenth clusters.

The seventh cluster features falling of petals in the courtyard. Based on the architectural style, this design plan takes up modern design techniques to stress clear lines in design and consistence with the exterior facade. The design still pursues the abundant and various natural atmosphere, harmony, plants landscape. The "falling" describes well the essence of petals in the courtyard—an elegant garden landscape like a young lady from respectable family, natural and graceful.

The ninth cluster features floating shadow and secret fragrance. In planning design of the ninth cluster, the sunny grassland with mottle tree shadows and secret fragrance form a grand landscape picture while the warm sunshine becomes the best camera of heart—everywhere is a picture and everywhere is a bright landscape line. The word of shadow implies a low-profile beauty, with no less charming and moving. As a song goes, "when petals leave the flower, there still remains secret fragrance". The word secret fragrance points out the low-profile state, graceful sense and luxuriant composed manner that the architects want to demonstrate.

The tenth cluster features fluttering butterflies following flowers. Using a pair of tenacious eyes to record the surrounding world through a Kaleidoscope and to collect all the happiness in a package, which is the central theme the tenth cluster wants to express. If eyes could be assimilated to a Kaleidoscope, then the tenth cluster wants to show a scene of color and happiness to you. The concept highlights the word of fluttering to speak the colorful landscape and happiness, full of passion and warm, which offer residents a final and satisfied home.

在细节设计上，项目以点带线，以线带面，注重景观的整体性；同时以绿化为重点，加强植物对景观效果及空间的营造；此外还非常注重景观的可参与性与可实施性，尽可能地在保证第七、九、十组团景观质量的前提下，降低设计成本。

第七组团方案特色——落，芳满庭。第七组团的方案设计是基于建筑的风格，采用现代设计手法，强调直线清晰、简洁明了的线形美感，与建筑外立面线形相统一。设计风格依然追求自然、和谐、植物造景形式的丰富和多样，是第七组团阐述景观的主要途径。一个"落"字道出了"芳满庭"恬静与从容的最终要义，设计所表现的是一个落落大方、宛如大家闺秀般的秀丽园林景观。

第九组团方案特色——影，浮暗香。在第九组团的方案设计中，如果说树影斑驳、暗香浮动的阳光草坪是一幅巨大的风景画，那么暖暖的阳光便是一部最好的心灵相机，穿越之处都是信手拈来的照片，处处都是一道道亮丽的风景线，瞄准一个"影"字，到处都是"低调的华丽"，又不失妩媚、动人之处。"当花瓣离开花朵，暗香残留"，"暗香"一词道出了设计师要表现的低调、从容得宛如顶尖大师所散发出来的理性与华丽自若。

第十组团方案特色——飘，蝶恋花。用万花筒看世界，固执地用眼睛记录周边的一切，贪婪地想把所有美好收起来。这就是第十组团的方案设计中所要表达的思想。如果说眼睛是一个万花筒，那么，第十组团想要带给人的就是满目色彩、满心欢喜的场景，其设计宗旨突出一个"飘"字，让纷纷扬扬的蝶恋花道尽万花筒般缤纷的风景与美好，充满了热情与温馨，让居者寻找到最终的归属与满足。

MODERN STYLE
现代风格

KEY WORDS 关键词

ELEMENT OF EMERALD
翡翠元素

VIEW SPACE
观景空间

LANDSCAPE NODE
景观节点

Location: Hangzhou, Zhejiang
Developer: Gemdale Corporation
Landscape Design: Shenzhen DongDa Landscape Design CO.LTD
Area: 88,000 m²

项目地点：浙江省杭州市
开 发 商：金地地产
景观设计：深圳市东大景观设计有限公司
面　　积：88 000 m²

Hangzhou Gemdale Zizai Town
杭州金地自在城

FEATURES 项目亮点

With emerald as the theme element, it uses a series of landscape groups and landscape architectures to create a modern, recreational and comfortable environment.

设计以翡翠作为主题元素，运用一系列的景观组团和景观小品打造了一个现代、休闲、舒适的景观环境。

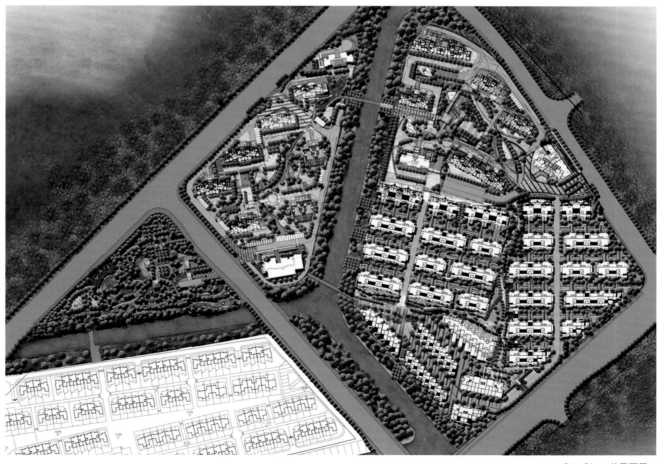

Site Plan 总平面图

Overview 项目概况

The project is located in Sandun Town, Xihu District of Hangzhou, adjacent to the municipal landscape river with flavored landscape ecology. With the maximized use of the location advantage, the community occupies abundant greening and various groups of waterscape for delicate decoration, which echoes the water body of municipal rivers. The design and construction of hard articles are exquisite to offer the elegant and comfortable atmosphere of high-quality dwelling.

项目位于杭州市西湖区三墩镇，濒临市政景观河道，环境生态舒适，设计上最大限度地利用了地区优势，小区内绿化丰富繁多，绿意盎然，同时设有多组水景精心点缀，与市政河道水体内外呼应，并在硬质小品设计施工上精雕细琢，营造出典雅舒适的高品质生活氛围。

Design Concept and Landscape Node 设计理念及景观节点

With emerald implying preciousness and auspiciousness as the design concept, three groups of themed landscape have been set for the project: Emerald Island, Emerald Necklace and Emerald Garden. As the coral landscape zone of the community, Emerald Garden has Natural Ecological Green Island as its center surrounded by scattering isle spaces, offering multiple dwelling experiences of distinct levels. Emerald Necklace Ecological Garden takes woods and streams as the humanity and life background to highlight the greening dwelling culture in Hangzhou. Green belts take the lead of landscape rhythm, connecting the landscape pieces by in curve. Emerald Garden group in the townhouse community is built with city flower of Hangzhou—May flower and magnolia to create the dwelling atmosphere of idyllic scenes, making itself a space of fitness, leisure and recreation for the elderly and kids.

设计理念以寓意珍贵吉祥的翡翠为元素，设翡翠岛、翡翠项链及翡翠花园等三组主题景观。其中翡翠岛花园为小区的核心景观区，以自然生态绿岛为中心，外围散布小岛式的观景空间，带来层次鲜明的多样化生活体验。翡翠项链生态花园则以树林水涧作为人文及生活背景，突出杭州的绿色居住文化。绿带引导景观节奏，以曲线串联各景观片段。组团翡翠花园位于联排小区，以杭州市花——桂花、玉兰建园，营造鸟语花香的生活气息，并结合为长者和儿童，以及健身设置休闲舒展的活动空间。

MODERN STYLE
现代风格

KEY WORDS 关键词

COURTYARD SPACE
庭院空间

OPEN VIEW
开敞视线

LANDSCAPE NODE
景观节点

Location: Nantong, Jiangsu
Developer: Nantong Wantong Properties Co., Ltd.
Landscape Design: Shenzhen DongDa Landscape Design CO.LTD
Land Area: 75,301 m²
Landscape Area: 60,200 m²
Plot Ratio: 2.056

项目地点：江苏省南通市
开 发 商：南通万通置业有限责任公司
景观设计：深圳市东大景观设计有限公司
占地面积：75 301 m²
景观面积：60 200 m²
容 积 率：2.056

Nantong One House Mansion
南通万濠华府

FEATURES 项目亮点

The landscape design focuses on green and emphasizes levels. Modern landscapes combine with the art-deco architectures perfectly.

景观设计以绿为主，强调层次感，运用现代的造园手法与装饰艺术风格的建筑进行融合，特色鲜明。

Site Plan 总平面图

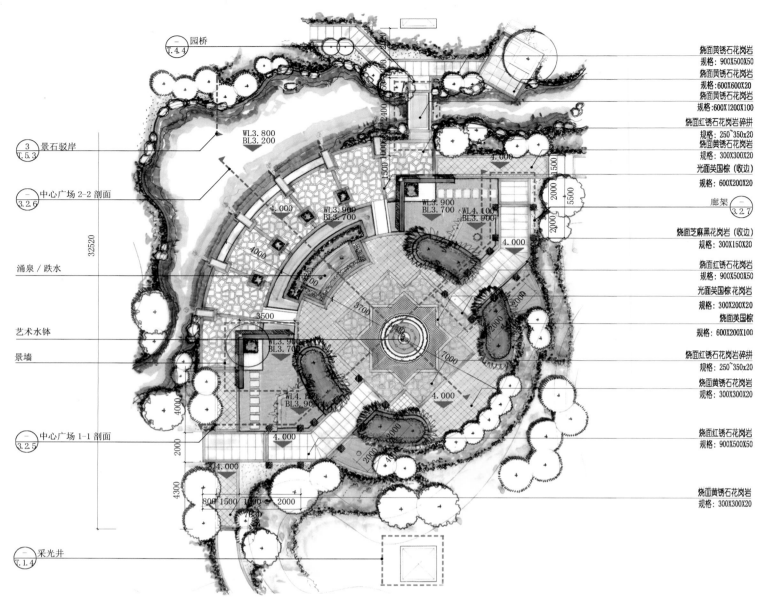

Plane of Central Plaza 中心广场平面详图

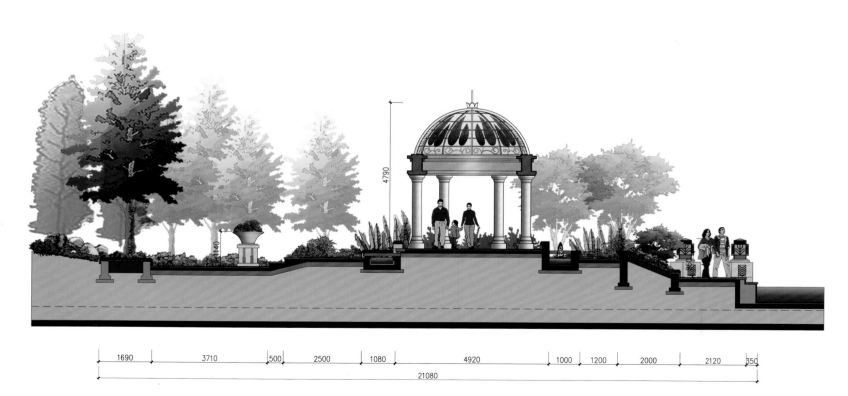

1-1 Section and Elevations of Pavilion Plaza
园亭广场 1-1 剖立面图

▶ Overview 项目概况

The project is located in downtown Nantong, at the intersection of Rengang Road and Haierxiang Road. Buildings are arranged in symmetrical layout, forming the east and west landscape courtyards. The buildings are designed with warm colors with some dark brown materials.

　　项目位于南通市城区，任港路与孩儿巷路交界处。建筑以中轴式对称布局，形成东、西两大主要景观庭园。小区建筑色彩以暖色为主导，搭配稳重的深咖啡色材料。

▶ Design Concept 设计理念

With the idea of "restoring the courtyard space and creating comfortable home", it applies modern art-deco design and pays attention to practical functions, bringing infinite imagination with changeful spaces.

　　以还原庭院空间、营造亲切家园为设计理念，运用现代的装饰艺术设计风格，注重质朴大方的语言和实用功能，采用变化的空间模式引发无限的空间遐想。

▶ Landscape Node 景观节点

Green landscapes of different levels are applied to create a comfortable living environment. In the east zone, the noisy activity center is set in the heart of the courtyard. Different activity spaces extend along the river to create fresh and pleasant waterfront environment. In the west zone, inside the spacious courtyard there is a large lake. Along the southeast bank, a wooden path is built to provide people with open views and water experience spaces.

整体设计以绿色为主，以大量有层次的绿化来营造舒适的人居环境。东区，将喧闹的活动场地安置在庭园的中部，各类型活动场地沿水溪逐渐展开，形成清新宜人的亲水环境。西区，利用宽阔的庭园空间打造开阔的大水面，沿水岸东南侧建造一条木栈道，形成开敞的视线空间，为居民提供精致舒适的亲水休闲活动空间。

MODERN STYLE
现代风格

KEY WORDS 关键词

WATERSIDE LANDSCAPE
水乡情韵

LANDSCAPE ELEMENT
景观元素

LANDSCAPE NODE
景观节点

Location: Qingpu District, Shanghai
Developer: Shanghai Gemdale Group
Landscape Design: Shenzhen DongDa Landscape Design CO.LTD
Area: 120,000 m²

项目地点：上海市青浦区
开　发　商：上海金地集团
景观设计：深圳市东大景观设计有限公司
面　　　积：120 000 m²

Gemdale Green County, Shanghai
上海金地格林郡

FEATURES 项目亮点

With "water" as the theme, the design applies bridges, streams and waterfront platforms as the main elements to emphasize the water-town style.

水是本项目景观设计的核心内容，方案中将"小桥、溪流、平台"作为重要的景观元素来审视，水被创造成为环境的焦点，彰显出水乡情韵的情怀。

Landscape Element 景观元素

Water is the theme for the landscape design. Bridges, streams and waterfront platforms become important landscape elements to provide water views for all the houses. Landscape architectures, walls and sites with simple and clear lines are designed around waterscapes, complementing the macro-environment. Simple and elegant design brings people with relaxing and pleasant experience.

　　水是本项目景观设计的核心内容，方案中将"小桥、溪流、平台"作为重要的景观元素来审视，使得家家有水景，水从楼前过，房后有花园。水被创造成为环境的焦点，彰显水乡情韵；而以清晰和简洁线条为特点构成的园建、景墙和场地则轻松地点缀在水景旁，与大环境"相得益彰"，简约而不简单，给人以轻松愉悦之感。

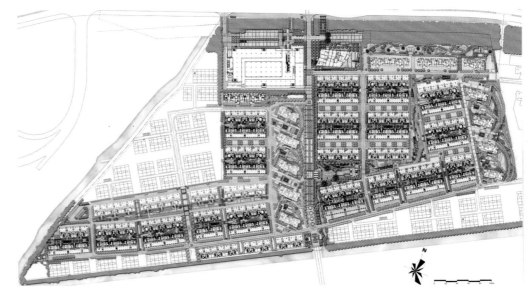

Site Plan 总平面图

MODERN STYLE
现代风格

KEY WORDS 关键词

ECOLOGICAL NATURAL
生态自然

LANDSCAPE DETAIL
景观细节

LANDSCAPE NODE
景观节点

Location: Hangzhou, Zhejiang
Developer: Hangzhou Xinlian Real Estate Co., Ltd.
Landscape Design: A&I International
Designers: Zhao Difeng, Xu Yang, Tong Liang
Land Area: 50,000 m²
Total Floor Area: 100,000 m²

项目地点：浙江省杭州市
开 发 商：杭州新联置业有限公司
景观设计：安道国际
设计人员：赵涤烽、徐扬、童亮
占地面积：50 000 m²
总建筑面积：约 100 000 m²

Hangzhou Lianhe Geli
杭州联合格里

FEATURES 项目亮点

This design creates a more unique and more delicate landscape space through dealing with details, gives abundant connotation and interest to landscape.

设计通过细节的处理来营造更独特、更精致的景观空间，赋予景观丰富的内涵和情趣。

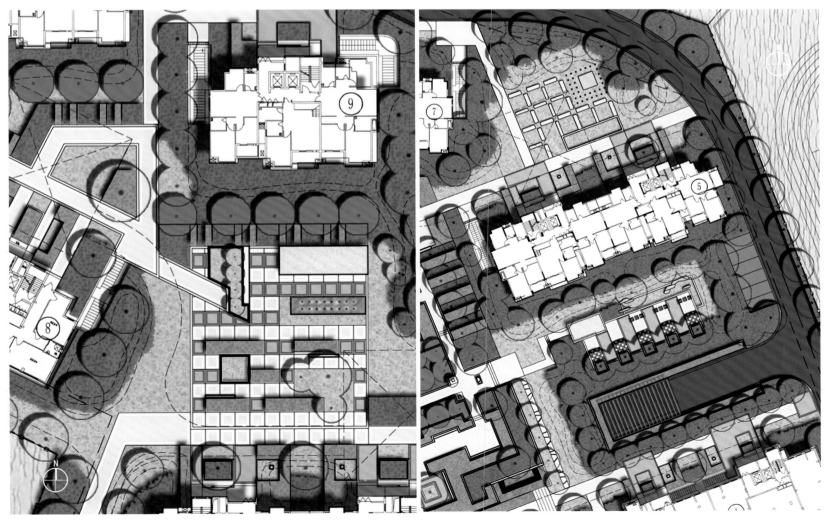

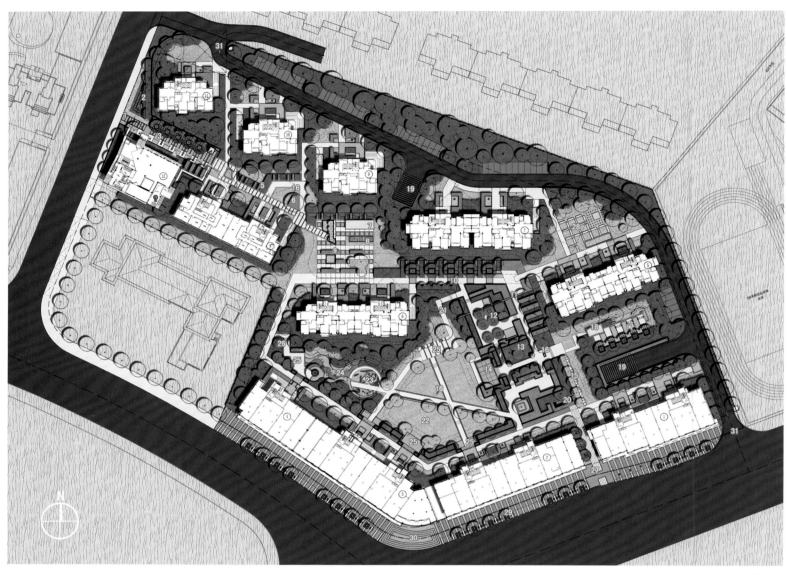

Site Plan 总平面图

Overview 项目概况

Located at CLD—Dingqiao Core Area, the project begins in the south from Danonggang Road and ends at Qinfeng Road in the west. Adjacent to Changmu Kindergarten and Changmu Primary School, it boasts convenient traffic.

联合格里位于杭州市主城区CLD——丁桥核心区，南至大农港路，西至勤丰路，倚靠长睦幼儿园、长睦小学，交通便利。

Style Positioning of the Project 风格定位

Seeking the minimalism of the essence of life. Minimalism is the constant phenomenon that the artists mentioned in the past few years and a way of thinking in fashion and contemporary art. Its feature is simplifying the elements, colors, lighting and raw materials in design to the utmost with the highest demand.

寻找生活本质的"简约主义"。简约主义是艺术家们在过去的几年中不断提及的现象，并应用在时尚及当代艺术中的一种思想方法。其特色是将设计的元素、色彩、照明、原材料简化到最少的程度，同时要求也是非常高的。

▶ Design Concept 设计理念

Light, planning and organic living. With transparent construction and sunny space, it presents a series of ecological, healthy, interesting landscapes. In Geli, you can be completely relaxed, adjust the mental space to get rid of cumbersome and pursuit simpleness truly.

光和规划、有机生活。设计秉承建筑的通透性与多阳光空间，景观呈现给人们的是生态的、健康的、有趣的。在联合格里，人们能得到彻底放松，调节精神空间，摆脱繁琐、复杂，真正地追求简单和自然。

▶ Design Approach 造景手法

Less is more. It endows rich connotation to simple things, through the process of details to create a landscape space with more special design and exquisite work, which would send out much more vitality and taste.

以少胜多、以简胜繁。通过简单来表现丰富要比借助于复杂来表现困难得多，这意味着需赋予简单的东西以丰富的内涵。通过细节的处理来营造一个更具有独特设计和精致做工的景观空间，因为细节的表现能赋予设计更多的生命和情趣。

MODERN STYLE
现代风格

KEY WORDS 关键词

ECOLOGICAL SPECIALTIES
生态特质

PICTURESQUE SCENERIES EVERYWHERE
移步异景

LANDSCAPE NODE
景观节点

Location: Changsha, Hunan
Developer: Vanke Changsha Lingyu Real Estate Development Co., Ltd.
Landscape Design: Shenzhen ALT Landscape Engineering and Design Co., Ltd.

项目地点：湖南省长沙市
开 发 商：万科长沙市领域房地产开发有限公司
景观设计：深圳雅蓝图景观工程设计有限公司

Changsha Vanke Golden Mansion Phase One and Two
长沙万科金域华府一、二期

FEATURES 项目亮点

The designers take full consideration of the coordination with surrounding architectures and interaction with urban environment. Various different designs were adopted, which achieved a three-dimensional, multi-level, green and comfortable space.

设计师充分考虑与周边建筑物的协调及城市的互动，采用多种设计手法，形成立体的、多层次的生态、绿色、舒适空间。

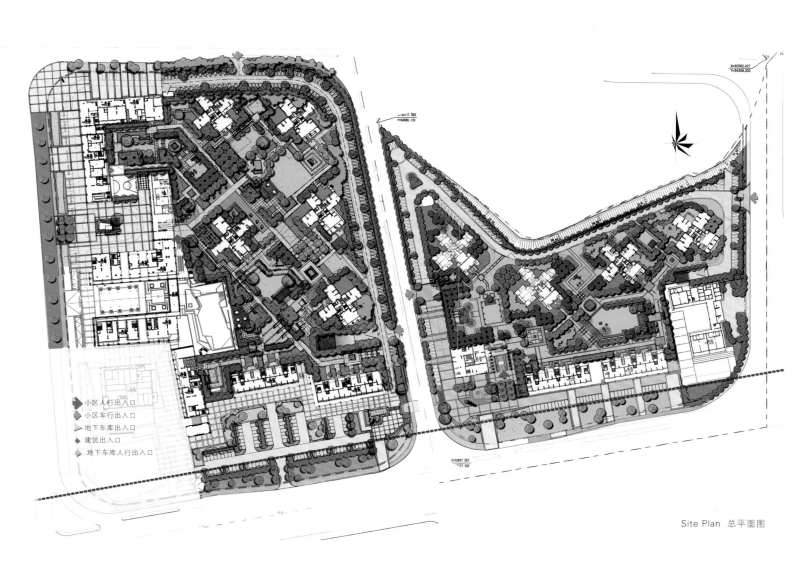

Site Plan 总平面图

▶ Overview 项目概况

Changsha Vanke Golden Mansion is located in the Second Class area of Changsha, Yuhua District, and central area of Wuguang New Town. It's sitting outside the second ring, with Xiangzhang East Road—main urban line on the south, Wanjiali Road—the main south-north line on the west and Hunan Unicom Headquarter on the north. The site is also closely next to District Government of Yuhua and Yuhua People's Square.

长沙万科金域华府项目位于长沙市Ⅱ类区——长沙市雨花区,武广新城的核心区域,二环外。地块南侧为城市干道香樟东路,紧临雨花区政府和雨花市民广场,西侧为长沙市南北主干道的万家丽路,北靠湖南联通总部。

▶ Design Concept 设计理念

Phase 2 applied modern design technique, which corresponds with landscape design of Phase 1 and takes full consideration of the coordination with surrounding architectures and interaction with urban environment. Various different designs were adopted, which would present a three-dimensional, multi-level, green and comfortable space after completion. People oriented landscape concept directed architects to put residents' experience at the first place through enclosing of different materials, design of life pieces and creation of landscape atmosphere. Meanwhile, architects not only meant to endow the space with experience and understanding on how residents feel, but also meant to strengthen cognition and communication to space.

项目采用了现代风格的多种设计手法,充分考虑了与周边建筑物的协调及城市的互动,项目建成后将形成立体的、多层次的生态、绿色舒适空间。以人为本的景观理念指导着设计师始终将不同的人们对景观的感受放在设计的首位,并且设计师希望赋予空间的不只是感觉上的体验和理解,而是思想对空间的认知和沟通。

▶ Landscape Node 景观节点

Landscape at the main entrance: the gateway of the residence. Here features water drop landscape, characteristic floor and the waterscape curtain wall surrounding the pool to create a tranquil, comfortable but luxury residential space opposite with the outside commercial space.

Between home landscape: Every family gets access to landscape; landscape changes with moving of steps. The resident in community needs not only open green land, but also private space which belongs only to oneself. The proper utilization of outdoor sports ground and overhead space, a combination of tranquil and dynamic, provides both hard and soft ground with full space and complete facilities to attract residents to walk out of rooms and join the public. Collocation of plants, walls and sketch facilities and treatment of leisure and activity space not only enhance the comfort of view, but also meet residents' requirement for safety and privacy.

Central garden: the guest-receiving room of the community. The central garden design of Phase 1 joins the water drop landscape at the entrance to form a centralized system. The perfect combination of dynamic drop and tranquil water and the proper arrangement of spring sculpture and landscape sculpture strengthen comparisons and changes. Along the water sets close water wooden platform—specialty sightseeing pavilion, which provides residents a mutual interactive space. The open step-like recreational grassland square links the central public activity area with private leisure space.

小区主入口景观：小区"门户"。通过电梯进入到小区，此处设计了跌水景观和入口特色铺地，以及泳池外侧的水景幕墙，营造出一种与外部繁华商业空间截然相反的静谧舒适且豪华大气的居住空间。

宅间景观：户户临景，移步异景。既需要开放的绿地，也需要有属于个人的私密空间。室外运动场所以及架空层空间的合理利用、动静结合，为住户提供面积充足、设施齐备的硬质及软质场地，以吸引用户走出房屋，参加公共活动，以增进住户间的交往。同时通过对植物、围墙和小品设施的配置和休闲活动空间景观处理，不仅能够增进视觉舒适度，也满足了居民对安全感和私密性的要求。

中心花园景观：小区"客厅"。一期中心花园景观设计延续入口跌水景观水系，形成一个集中水域。动态跌水与平静水面的完美结合，喷水雕塑与景观雕塑的合理布局，增强对景的同时又富于变化，沿水区有亲水木平台，特色景观亭更为住户创造了一个与水互动的空间。开放性台阶式休闲草坪广场成为中心公共活动空间和私密休闲空间之间的自然过渡。

Landscape Detail 景观细部

Frame shaping: different qualities of lines are applied to form regular collision and interweaving. The independent surfaces generated from the collision of lines fulfill and enrich the entire frame structure through the figure-ground relations of plants, floor and water. In searching of artistic inspiration, functional appropriateness and comfort were also taken into consideration as well for the purpose of "searching for life in art and experiencing art in life".

Exquisite details: the application of modern decorative elements such as glass, metal, wood and stones in details is another view and experience the design wants to express to people. Through the collocation of these materials and exquisite sizes, a simple, concise but elegant landscape temperament has been forged.

　　框架的形成：通过运用不同粗细的艺术线条，有规律地进行碰撞和交织。在线与线交汇碰撞中形成的面与面之间通过植物、铺装、水体等不同的图底关系，丰富和充实着整个框架结构。在寻找艺术灵感的同时也考虑到功能上的合理性和使用上的舒适性，力求在艺术中寻找生活，在生活中感受艺术。

　　细节的考究：玻璃、金属、木材、石材等现代的装饰元素在细节上的运用是设计所要表达的另一种视觉和触觉的体验。通过这些材料的搭配和尺寸的考究，营造出简约、朴实又不失高贵典雅的景观气质。

MODERN STYLE
现代风格

KEY WORDS 关键词

- ECOLOGICAL RESOURCE 生态资源
- PHYTOCOENOSIUM 植物群落
- LANDSCAPE NODE 景观节点

Location: Zhuhai, Guangdong
Developer: Zhuhai Zobon Real Estate Development Co., Ltd.
Landscape Design: L&A Design Group
Site Area: 147,653 m²

项目地点：广东省珠海市
发 展 商：珠海中邦房地产开发有限公司
景观设计：奥雅设计集团
基地总面积：147 653 m²

ZOBON City Villa, Zhuhai
珠海中邦城市美墅

FEATURES 项目亮点

It takes great advantage of the river view, and sets planting groups according to the space ambiances to show the beauty of four seasons.

项目最大限度地引入场地内河道的自然风光，结合不同区域的空间氛围创造特色鲜明的植物群落，展现四季流转的变化之美。

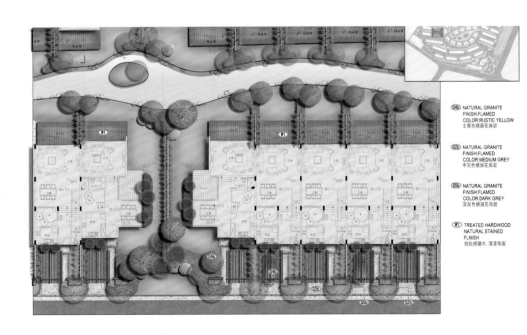

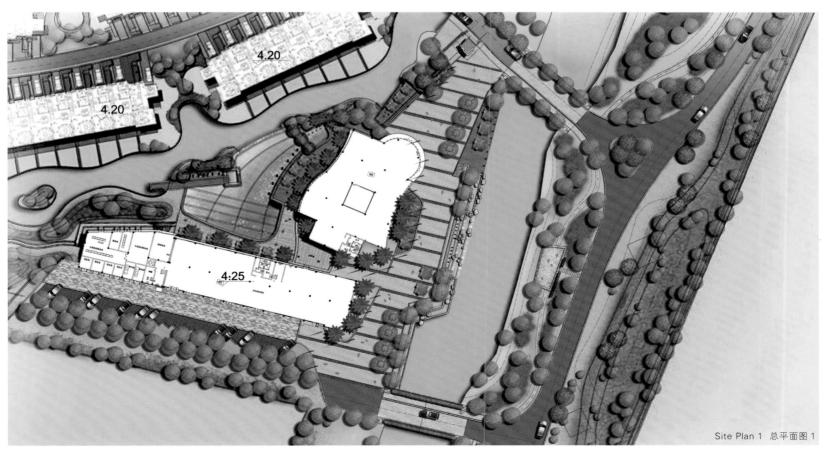

Site Plan 1 总平面图 1

Site Plan 2 总平面图 2

Overview 项目概况

The project is located on the north of Shining River Front Garden, Baijiao Town of Doumen District, Zhuhai, with a river on the north, a flood-drainage channel on the east, and a 12 m and 24 m wide road on the south.

本项目位于珠海市斗门区白蕉镇金碧丽江花园北侧。用地北面为天然河道，东面现状为排洪渠，用地南面分别与 12 m 道路和 24 m 道路相邻。

Design Concept 设计理念

Following the idea of "islands and spiral" in architectural planning, brooks are designed to connect the building clusters and provide all the families with their own bays.

该项目顺应建筑规划"群岛和螺旋"的设计理念，用溪流贯穿联系各大组团板块，最大限度地利用生态水资源，让每个家庭拥有一个自己的水湾。

Landscape Design 景观设计

The designers aim for a modern high-end residential community with pure waterfront garden.

Home—warm and convenient. It tries to create reasonable functional areas, well-organized flowing lines, convenient traffic system, and caring supporting facilities. All of the demands of the users' are carefully considered. It is not only a residence but also a home for heart bearing happiness and hope.

Garden—natural and healthy. The design tries to integrate life into nature which is an ideal for living. It takes great advantage of the river view, and sets planting groups according to the space ambiances to show the beauty of four seasons. Residents will not only get pleasant visual experience but also relax themselves to regain the peace of heart.

设计师致力于打造一个现代自然的纯水岸花园的综合性高档社区。

家园——温馨便捷。项目设计定位是打造最合理的功能与流线组织、最便利的交通路线、最贴心的配套设施，为每一位使用者的生活需求作细致的考虑。它提供的不仅仅是一个居所，而是一个承载幸福与希望的"心"的家园。

花园——自然健康。设计力求缔造与自然融为一体的生活方式，这也许是人类居住的最高境界。该项目设计最大限度地引入场地内河道的自然风光，结合不同区域的空间氛围创造特色鲜明的植物群落，展现四季流转的变化之美。它提供的不仅仅是视觉上的愉悦享受，更使身居其中的人在绿色包围中消解身心疲意，重获发自内心的宁静安详。

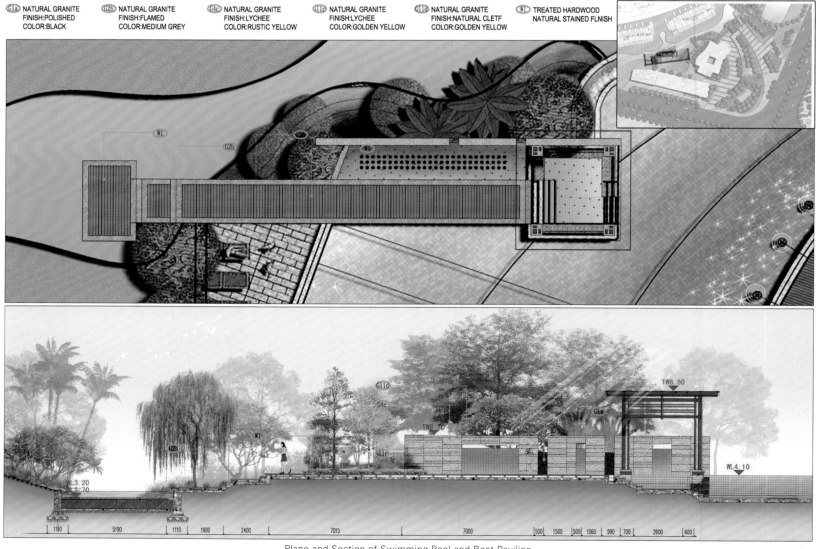

Plane and Section of Swimming Pool and Rest Pavilion
泳池休息亭平面图、剖面图

South Elevation of Main Entrance
主入口南立面图

North Elevation of Main Entrance
主入口北立面图

W1 TREATED HARDWOOD NATURAL STAINED FLNISH　　GL1 TEMPERED GLASS ROOFING　　G2b NATURAL GRANITE FINISH:FLAMED COLOR:MEDIUM GREY　　ST1 PRE-FABRICATED STAINLESS STEEL

Elevation for Type 1 Unit
典型户型一立面图

Neoclassical Style
新古典主义风格

Color Design
强调色彩

Elegant Effect
典雅效果

Double Style
双重气质

NEOCLASSICAL STYLE
新古典主义风格

KEY WORDS 关键词

ECOLOGICAL LANDSCAPE
生态景观

DIVERSE SPACES
多样空间

LANDSCAPE NODE
景观节点

Location: Jiangyin, JiangSu
Developer: Jiangyin Jinke Property Development Co., Ltd.
Landscape Design: Shenzhen Art-spring Landscape Design Co., Ltd.
Landscape Area: 75,000 m²

项目地点：江苏省江阴市
开 发 商：江阴金科置业发展有限公司
景观设计：深圳市阿特森泛华环境艺术设计有限公司
景观面积：75 000 m²

Jinke·Oriental Palace
金科·东方王府

FEATURES 项目亮点

Neo-classicism is sufficiently considered in the landscape design, expressing the yearning for nature with the ecological verdurization surroundings.

项目的景观设计充分地考虑了建筑的新古典主义风格，用一种生态绿化包围的方式来表达对自然的向往。

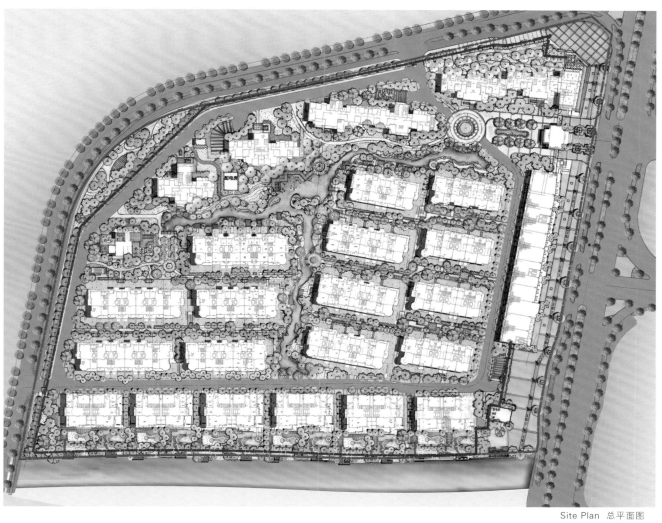

Site Plan 总平面图

Design Concept 设计理念

The project is a combination of neo-classicism and ecological landscape. Neo-classicism inherits from the traditional architectural aesthetics but eliminates the heavy complication of classicism; it is not simple and cold as modernism, either. It discards the weaknesses of classicism and modernism, but absorbs their advantages. The artistic deposits of classicism and the functionality of modernism are combined perfectly, creating the beauty in formal in the post-industrial age and modern living style.

Neo-classicism is the main tone for the design, integrated into the nature with neo-classicism skill and elements to display the dignity and quality of the project. The ecological verdurization surroundings express people's yearning for nature, while the new Art Deco defines the dignity and elegance. The verdurization surrounds the buildings; the buildings grow in the verdurization; the landscape and the nature are integrated.

项目采用新古典主义与生态景观的对话，新古典承袭传统建筑美学，没有古典主义的繁复复杂，也不似现代主义过于简短的冰冷而缺乏亲近感，新古典主义建筑避免了它们的缺点，却融合了两者的优点，将古典主义的艺术底蕴和现代主义的功能性完美结合，创造了后工业时代一种有厚度的形式美，给予了现代人居绝佳的答案。

设计将新古典主义的风格主基调融入自然，以新古典主义手法与元素来体现项目的尊贵感与品质感。以一种绿化包围的方式表达人们对自然的向往，新装饰艺术的奢华则诠释了尊贵与典雅，让绿化包围建筑，让建筑生存在绿化之中，让景观设计与自然环境融合。

NEOCLASSICAL STYLE
新古典主义风格

KEY WORDS 关键词

ECOLOGICAL CHARACTERISTICS
生态特质

NATURAL VERDURIZATION
自然绿化

LANDSCAPE NODE
景观节点

Location: Guilin, Guangxi
Developer: Guilin Songyu Investment Co., Ltd.
Landscape Design: Shenzhen Art-spring Landscape Design Co., Ltd.
Landscape Area: 55,000 m²

项目地点：广西壮族自治区桂林市
开 发 商：桂林市松煜投资有限公司
景观设计：深圳市阿特森泛华环境艺术设计有限公司
景观面积：55 000 m²

Xingjin · Mansion of Shang 兴进·上郡

FEATURES 项目亮点

Neo-classicism is the main tone for landscape design, integrated into the nature and providing a perfect living space for the modern classy consumers.

景观设计将新古典主义风格的主基调融入自然，给现代高端消费者提供一个居住绝佳之境。

Overview 项目概况

Based on the knowledge on architectural design, analysis on the site condition and the understanding to the target consumers, the landscape style of this project is positioned as neo-classicism with characteristics of dignity, steady and elegance.

基于对项目建筑设计的解读，基地现状及周边楼盘的分析，以及设定目标消费群体的了解，兴进·上郡项目景观风格定位为新古典主义风格。以尊贵、稳重、典雅定位该项目景观的品质。

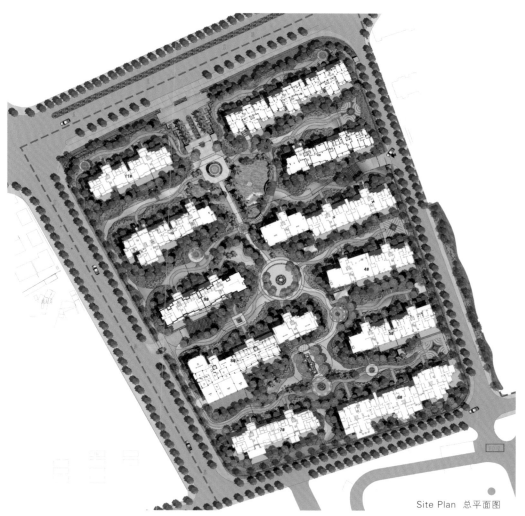

Site Plan 总平面图

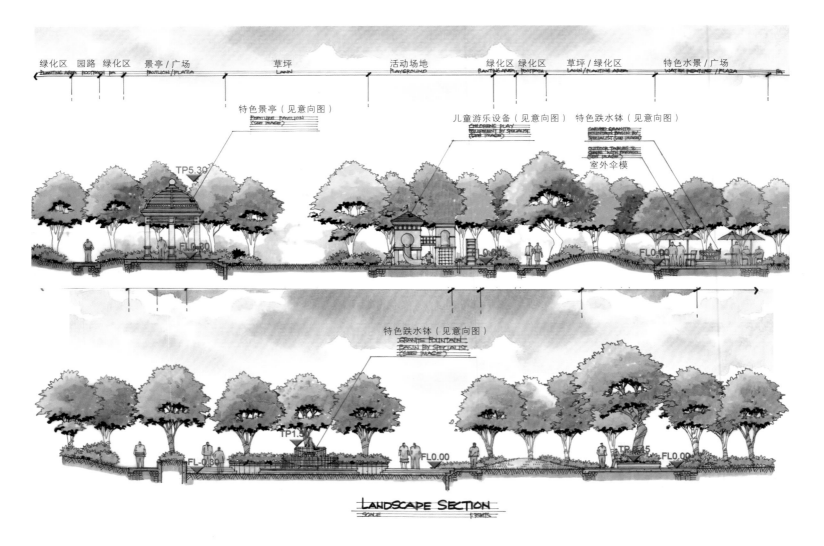

▶ Design Concept 设计理念

Neo-classicism inherits from the traditional architectural aesthetics but eliminates the heavy complication of classicism; it is not simple and cold as modernism, either. It discards the weaknesses of classicism and modernism, but absorbs their advantages. The artistic deposits of classicism and the functionality of modernism are combined perfectly, creating the beauty in formal in the post-industrial age and modern living style.

新古典主义是承袭传统建筑美学，舍去古典主义的繁复不堪，也不似现代主义过于简单的冰冷而缺乏亲近感，新古典主义建筑避免了两者的缺点，却融合了两者的优点，它将古典主义的艺术底蕴和现代主义的功能性完美结合，创造了后工业时代一种有厚度的形式美，给予了现代人居的绝佳答案。

▶ Landscape Node 景观节点

Neo-classicism is the main tone for the landscape design for this project, integrated into the nature with neo-classicism skill and elements to display the dignity and quality of the project, and with natural verdurization to show the ecological, natural and healthy living concept. They are combined perfectly and benefit each other, building the living atmosphere of neo-classicism, park and home and providing a perfect living space for the modern classy consumers.

兴进·上郡项目景观设计以新古典主义风格的主基调融入自然，以新古典主义手法与元素来体现项目的尊贵感与品质感，以自然绿化体现生态、自然、健康的生活理念，两者完美结合，共生共融，营造新古典主义、公园、家的生活氛围，给现代高端消费者提供一个居住绝佳之境。

NEOCLASSICAL STYLE
新古典主义风格

KEY WORDS 关键词

NATURAL ECOLOGY
自然生态

HUMANITY LANDSCAPE
人文景观

LANDSCAPE NODE
景观节点

Location: Zhongshan, Guangdong
Developer: Sino-ocean Land (Zhongshan)
Landscape Design: Shenzhen Art-spring Landscape Design Co.,Ltd.
Landscape Area: 25,000 m²

项目地点：广东省中山市
开 发 商：远洋地产（中山）开发有限公司
景观设计：深圳市阿特森泛华环境艺术设计有限公司
景观面积：25 000 m²

Ocean City, Zhongshan
中山远洋城

FEATURES 项目亮点

"Mountain" and "water" are the keywords of landscape design, which presents the integration of humanity design and natural ecological environment. It is the generality of the landscape design development of the project.

"山水"是项目景观设计的关键词，其体现了项目人文设计与自然生态环境的融合，是该项目景观设计发展的共性。

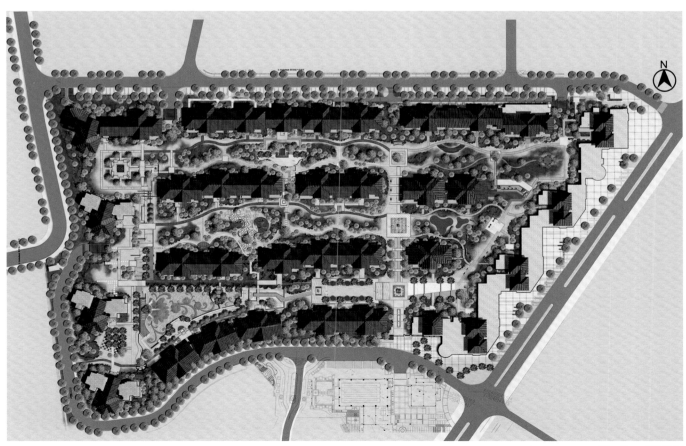

Site Plan 总平面图

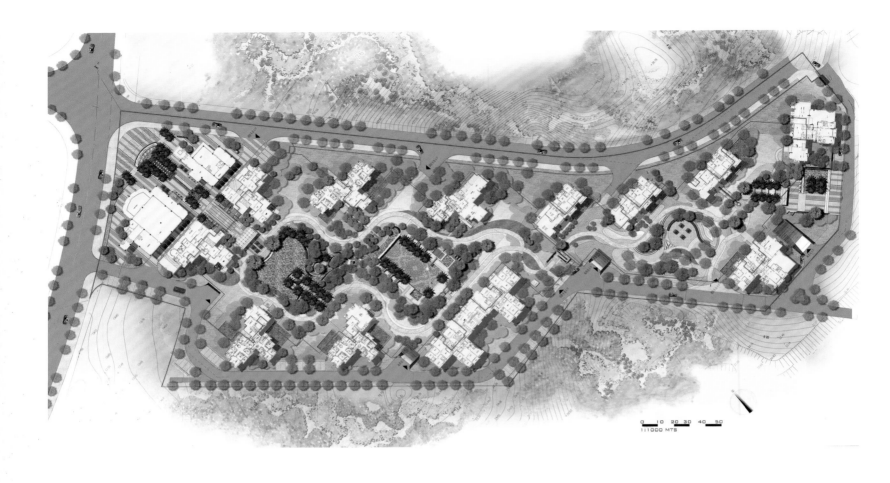

Overview 项目概况

As an international living place of high quality, the project is located in the center of the future downtown of Zhongshan – the ecological region made up of Zimaling South, the national Level A ecological protection zone Wugui Mount, Jinzi Mount and the three ecological zones of Zimaling. The total planned land area is about 1,000,005 m² and the total planed floor area is about 2,000,000 m². As the model of HOPSCA urban planning operation mode, the project will be built to be an international trade community consisting ecology, amusement, commerce and residence.

中山远洋城是远洋地产以其国际化视野精心打造的国际品质生活范本，雄踞中山市未来城市中央——紫马岭南片区，国家A级生态保护区五桂山、金字山、紫马岭三大生态区围合而成的中心生态腹地。总规划建设用地近1 000 005 m²，总规划建筑面积约2 000 000 m²。作为HOPSCA城市规划运营模式的典范，项目将建成集生态、休闲、商务、居住于一体的国际商务社区。

Design Concept 设计理念

According to the planning, the project was developed in several periods and designed with neo-classicism and simple classicism style, forming a harmony and integrated urban space. Based on the planned space and the feature of the sites, the product position and the surrounding relations, the design formed the exterior landscape spaces with differences. Integrated with the concept of "city of mountain and water", the project presents the feature of ecological region. "Mountain" and "water" are the keywords of landscape design, which presents the integration of humanity design and natural ecological environment. It is the generality of the landscape design development of the project.

中山远洋城以总体规划为指导依据，实施分期开发，并以新古典及简约古典风格作为项目建筑的设计风格，形成了协调与统一的城市空间界面。各期规划则以总体规划空间为依据，根据地块基地特征、建筑产品定位及周边关系形成具有差异化的外部景观空间。项目规划融入"山水之城"的设计理念，充分体现了项目生态腹地的基地特质。"山水"亦是项目景观设计的关键词，其体现了项目人文设计与自然生态环境的融合，是该项目景观设计发展的共性。

Landscape Node 景观节点

The site of the project extends from the sub-center of the city to the ecological park, while A4, A5 and A11 are the different spatial nodes on the disposition axis, which shows different features of landscape along the space. A4 and A5 are the connections of the city and the nature, presenting the exterior commercial public landscape and the interior green landscapes at the same time. The green park roads, the sunshine meadow, the curled paths, the small streams and the natural ponds are the paintings on the green base.

Among the small high-rise group, the natural gardens, other than the regular buildings, are meeting the demand for the small high-rise construction space. A11, however, presents another spatial atmosphere with roads around the mountain and the mountain as the background of the site. The site is also an extension of the mountain. The two large terraces display the atmosphere of mountain life. The space is open but the dotted high-rise buildings form the enclosure. By the distances between the buildings, the landscape outside is seen from the inside. Most of the buildings are not close but grow among the landscape.

According to the demand for spatial scale of the construction and the site, A12 is designed as landscape space of general forest park, extending the dropping relation from south to north and forming different nature and humanity landscapes from valley to forest lake to platform.

项目基地由城市副中心向生态山体公园延续，A4/A5及A11则是此序列轴上的不同空间节点，依据空间递进其景观又体现为不同的特质。A4/A5是城市与自然的交融点，其在体现外围商业公共景观界面的同时，内部住区主要体现了疏林坡地的景观绿色基底，林荫园路、刚劲的阳光草地、曲线的园径、折线跌宕的小溪、自然的池塘及镜面的翡翠泳池是绿色基底上的画。

在小高层建筑的群落中，体会的是自然花园，而不是规则的行列式建筑空间，这种尺度恰恰是小高层建筑空间的诉求。A11却是另外一种空间氛围，周边道路对山体的渗透与对景，山体是基地的背景，基地亦是山体的延伸，两个大型的台地关系已经让人体味到山居的氛围。这里的空间是开敞性的，四周的点式高层建筑给人以空间围合感，但通过建筑之间的间距，将外面的景观渗透进来，多数建筑不是封闭的界面，而是生长在景观之中。

考虑到A12建筑与场地空间比例的需求，将其设计为泛森林公园化的景观空间场景，由南侧（临近生态山体）向北，延续高差的跌落关系，由山谷—森林湖—镜面水台形成了绿与水的不同自然人文景观形态。

NEOCLASSICAL STYLE
新古典主义风格

Jiangsu Changzhou Future Golden County
江苏常州新城金郡

KEY WORDS 关键词

NATURAL ECOLOGY
自然生态

SLOPE LANDSCAPE
坡地景观

LANDSCAPE NODE
景观节点

FEATURES 项目亮点

This project is mainly romantic amorous feelings with neoclassical style and natural ecological landscape. A fresh and elegant field is created the enrichment and extraction of the cultural conditions, through " trees, waterscape, fresh breeze, blue sky, music", builds up a fresh and elegant field.

本案以新古典主义的浪漫风情和自然生态景观为主脉，通过"树林、水景、清风、蓝天、音乐"对文化风情的浓缩、提炼，营造出清新淡雅的国度。

Location: Changzhou, Jiangsu
Developer: Jiangsu Future Land Co., Ltd
Landscape Design: America EDSK International Design Group (China) Co., Ltd
Chief Designer: Sun Lihui, He Xin, Zheng Huiguo, Jin Nanbin
Land Area: 77,820 m²

项目地点：江苏省常州市
开发商：江苏新城地产集团
景观设计：美国 EDSK 易顿国际设计集团（中国）有限公司
主要设计人员：孙立辉 何鑫 郑辉国 金南斌
用地面积：77 820 m²

▶ Overview 项目概况

Future Golden County locates in the west of Hongmeinan Road, the north of Qingliangdong Road, and the south of Laodongdong Road, the total plot area of which is about 77,817 m². Being the good quality houses, Future Golden County successfully emerges neoclassical conception into landscape design, building up a community environment possessing landscape architecture with romantic amorous .

新城金郡位于江苏省常州市红梅南路以西,清凉东路以北,劳动东路以南,规划总用地约为 77 817 m²。作为常州品质栖息楼盘,新城金郡将新古典主义的意境融入景观设计中,营造了一种远离城市喧嚣的浪漫风情园林景观社区环境。

▶ Design Concept 设计理念

The landscape design concept comes from neoclassicism style, pursuing the variation of space, movement, and time. The detail design takes ecological nature as fundamental, respects nature, and creates theme spaces having different functions and different features. From the whole landscape design, the romantic amorous with neoclassical style and natural ecological landscape is the main vein, A fresh and elegant field is created the enrichment and extraction of the cultural conditions, through " trees, waterscape, fresh breeze, blue sky, music", builds up a fresh and elegant field.

景观设计的灵感源自新古典主义风格,在设计上追求空间、动静、时间上的变化等特色。具体设计以生态自然为根本,尊重自然,创造出不同功能、不同特性的主题空间。在整个园林设计中以新古典主义的浪漫风情和自然生态景观为主脉,通过"树林、水景、清风、蓝天、音乐"对文化风情的浓缩、提炼,营造出清新淡雅的国度。

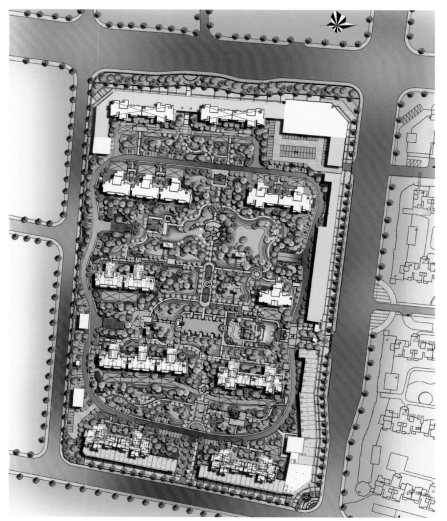

Site Plan 总平面图

Landscape Node 景观节点

The underground double-decked garage has been a feature of this houses, producing a distinctive slope landscape. In the natural landscape, water is a vital landscape element, making use of its altitude difference to create dynamic water-fall waterscape. Absorbing the essence of western garden making, the design concept is mainly point water spray, and part is supplemented by formal style waterscape, combining the European style water spray and wide grassy slope so as to create an ecology environment integrating dynamic and quietness.

The water-fall wall facing the entrance axis is the visual focus, approaches of opposite scenery are adopted so as to viewing a mirror-like waterscape. The home landscape is more than a leisure space considering the residence privacy. Each sight spot is full of poetic charms, thus, the residents feel like enjoying in a wonderful garden.

Highlighting the nature feelings is the foundation of the plant landscape design for this project, stressing on aspect change and being based on the layered collocation of trees, bushes and grass , utilizing variety plants to create a multifunctional garden space. While the public area stands out the safety facility for the aged and children activity. What is more, the protection for ground floor residents reflects the human-oriented philosophy.

Through bringing in the native trees from local forests, the greening rate is expected to be improved as much as possible. The spatial contrast variation between high density planting area and wide grass area feels like in the atmosphere of " wind path". In view of plant shapes and layering , besides foliage plants, there are plants with multiple uses, such as fruit trees, hanging plants and perennials as well as the trees using for attracting birds. As a result, a balance, stable and abundant community ecological food chain consists of " trees attract birds, birds catch worms, worms fertilize soil, soil cultivates trees", reducing anthropogenic pollution to the community environment and producing a sustainable development natural environment.

小区的双层地下车库是一大特色，可以形成极具特色的坡地景观。在自然园林中，水是极重要的景观要素，利用其高差关系，可以营造灵动的跌水景观。设计吸收西方造园的精华，以点式喷水为主、局部以规则式水面为辅的核心设计理念，将欧式喷水与开阔的草坡相结合，营造出一个动感与幽静相结合的生态环境。

入口轴线正对的跌水景墙是入口景观的视觉焦点，应用了造园中的对景手法，于入口处便可看到一平如镜的水景。宅间景观考虑到住户的私密性，设为休憩的空间。设计的每一处景点都充满了诗情画意，让住户体验到优美的园林享受。

项目的植物景观设计以突出自然风情为根本，强调季节变化。以乔、灌、草分层搭配为原则，运用品种多样的花草树木，形成功能多样的园林空间。公共组团区强调老人与儿童活动的安全设置，以及对首层住户私密性的保护，体现以人为本的设计理念。

在绿化方面引进当地自然山林的乡土树种，尽量提高绿化率。运用高密度的植栽区域与空旷的草坪区域的空间对比变化，使人感受到"曲径通幽"的氛围。考虑到植物造型及层次感，除了观叶植物外，还种植了果树、攀沿垂吊类植物和宿根类的草花以及诱鸟树等多种用途的植物，形成"树引鸟、鸟吃虫、虫肥土、土养树"的平衡、稳定、丰富的社区生态食物链，减少人为造成的社区环境污染，形成可持续发展的自然生态环境。

NEOCLASSICAL STYLE
新古典主义风格

KEY WORDS 关键词

ECOLOGICAL QUALITY
生态特质

CASCADE
跌水景观

LANDSCAPE NODE
景观节点

Location: Changzhou, Jiangsu
Developer: Jiangsu New Town Real Estate Group
Landscape Design: America EDSK International Design Group (China) Co., Ltd.
Chief Designer: Sun Lihui, He Xin, Zheng Huiguo, Wang Hangdong
Total Land Area: 185,080 m²

项目地点：江苏省常州市
开 发 商：江苏新城地产集团
景观设计：美国EDSK易顿国际设计集团（中国）有限公司
主要设计人员：孙立辉 何鑫 郑辉国 王航东
总用地面积：185 080 m²

Changzhou Xiangyi Zijun New Town
江苏常州新城香溢紫郡

FEATURES 项目亮点

Traditional French-style garden serves as the landscape axis in the central area. In addition with the small decorations and landscape nodes, it creates an integral landscape system of European style.

园林中心景观区以中轴线（法式古典花园）为中心进行布局，主题小品与主要景点穿插其间，形成一个完整的欧式古典景观体系。

▶ Overview 项目概况

The project locates in Qisuyan, Changzhou, Jiangsu with a total land area of 185,080 m², which is in the west of Qingyang Road, north of the Orient West Road, south of the Longcheng Highway. It is 2.6 km away from the Changzhou East Station of Hu-Ning high-speed rail and 4.5 km from the Changzhou railway station, and gets close to the elevated of Qingyang Road.

项目处于江苏省常州市戚墅堰区，青阳路西侧，东方西路北侧，龙城大道南侧，总用地面积 185 080 m²。地块距沪宁高铁常州东站直线距离 2.6 km，距常州火车站 4.5 km，紧邻青洋路高架桥，交通便捷。

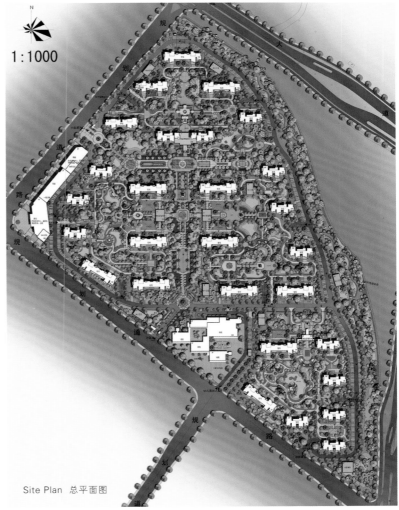

Site Plan 总平面图

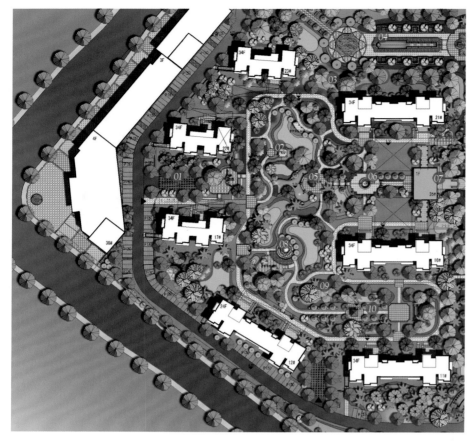

① 车库入口
② 休憩构架
③ 自然坡地
④ 规则水景
⑤ 中央景亭
⑥ 欧式跌水
⑦ 组团入口
⑧ 休憩廊架
⑨ 绿化空间
⑩ 小区园路

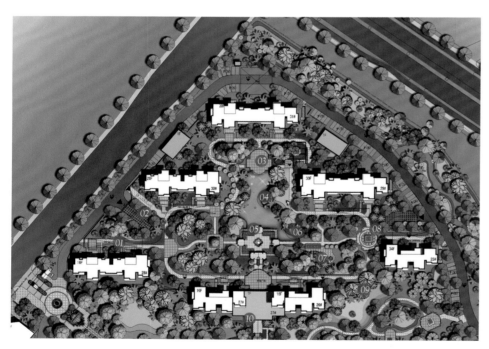

① 住户入口
② 车库入口
③ 休憩平台
④ 自然水系
⑤ 中心景亭
⑥ 绿化空间
⑦ 儿童游乐
⑧ 特色铺装
⑨ 坡地景观
⑩ 组团入口

① 绿化空间
② 组团入口
③ 对景景墙
④ 点式水景
⑤ 开阔草坪
⑥ 住户入口
⑦ 儿童游乐
⑧ 宅间绿化
⑨ 车库入口
⑩ 消防入口

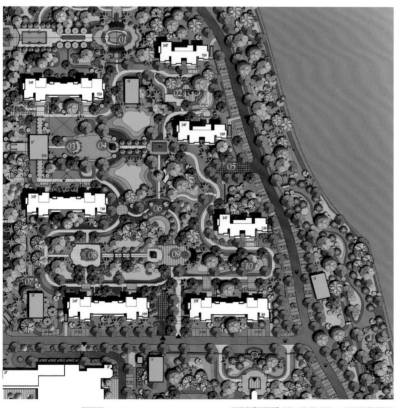

① 中心景观
② 儿童游乐
③ 特色铺装
④ 规则草坪
⑤ 车库入口
⑥ 老年活动
⑦ 自然坡地
⑧ 景观花坛
⑨ 中心景观
⑩ 绿化空间

Design Concept 规划理念

Two-axis, multifunctional, four group-landscapes, one greenbelt

Central "T" Axis: Horizontal landscape axis—entrance landscape area, putting-green-type golf landscape, boulevard, waterside footpath, water-fall landscape, small fitness plaza, leisure fitness footpath, sunshine lawn, etc. longitudinal landscape axis—leisure corridor, elegant landscape pond, formal lawn, etc.

Multifunctional—all kinds of activity space provide people with a comprehensive place of leisure, activity, fitness and entertainment.

Four group-landscapes—All kinds of group-landscapes and planning bring out the best in each other that perfectly combine the formal beauty of the landscape and the residential function, fully considering the residents' behavior custom.

Greenbelt in the east—The greenbelt is a natural cover for noise defense and functions as a "green lung" to add more "greens" for the community, establishing the sustainable development of the natural ecological environment.

"两轴、多点、四组团、一绿化带"。

中心"T"字轴：横向景观轴——入口景观区，果岭型高尔夫景观，林荫大道，中心水景步道，中心跌水景观，健身小广场，休闲健身步道，阳光草坪等。纵向景观轴——休闲走廊，雅致景观水池，规则式草坪等，形成多种活动休闲的空间，具有强烈的景观序列感。

多点——各组团的生活活动小空间，为居民提供了休闲、活动、健身、娱乐集于一体的生活空间。

四重主题组团景观——各具特色的组团景观，与规划相得益彰，将景观的形式美与居住功能性相结合，充分考虑居住者的行为习惯，同时赋予景观人文内涵。

东侧绿化带——项目东侧有市政高架，绿化带成为一道隔绝噪音污染的天然屏障，同时成为社区的绿肺，为小区多添加一份绿意，形成可持续发展的自然生态环境。

Landscape Feature 景观特色

The designers skillfully bring the reinforced concrete slab and the road into a superior eco habitat which has clear sky, green lands and various living beings. They not only build the shape but also infuse the buildings with the surroundings to create a new living style.

The project has 13 landscape areas: entrance landscape area, putting-green-type golf landscape, fog landscape, boulevard, waterside footpath, water-fall landscape, elegant landscape pond, leisure fitness footpath, sunshine lawn, sloping fields, residential buildings and Business Street.

The plants landscape lays stress on the ecology, hierarchy, interestingness and richness, thus creating both rich plant species and hierarchical green space. It is rooted in the natural feelings and appropriately goes on landscaping on the local mountain forest & vegetation to achieve the seasonable scenery.

"以人为本、天人合一"是本案所体现的宗旨。设计师科学地运用多种现代艺术处理手法，巧妙地将钢筋混凝土楼板及道路融合到一个有晴空、绿地、多样物种并存的优质的生态栖息环境当中。在这里，设计师不只是单纯地造型或注重气氛的营造，同时，让建筑与周围的环境结合并相融，从而让居住者产生新的生活方式与灵感。

项目共形成了13大景观节点，分别是：入口区域景观、果岭型高尔夫景观、雾森景观、林荫大道、中心水景步道、中心跌水景观、休闲健身步道、雅致景观水池、休闲走廊、规则式草坪景观、坡地景观、住宅间景观以及商业街景观，各节点之间环环相扣，从而形成了多层次的景观体系。

在植物的设计上强调景观植物的生态性、植物的层次性、植物的趣味性以及植物的丰富性，使得整个香溢紫郡不仅拥有丰富的植物树种，还富有多重层次的绿化空间；同时通过在重要的节点植物的营造，达到观赏的趣味性。植物景观设计以突出自然风情为根本，强调季节变化，并适当地运用当地的山林植被进行造景。

CIMC Unique Wenchang Landscape Design
扬州·中集紫金文昌景观设计

NEOCLASSICAL STYLE
新古典主义风格

KEY WORDS 关键词

ECOLOGICAL SPACE
生态空间

3D EXQUISITENESS
立体精致

LANDSCAPE NODE
景观节点

Location: Yangzhou, Jiangsu
Developer: Yangzhou CIMC Dayu Properties Co., Ltd.
Landscape Design: Shenzhen ALT Architectural Landscape Design Co., Ltd.

项目地点：江苏省扬州市
开 发 商：扬州市中集达宇置业有限公司
景观设计：深圳市雅蓝图景观工程设计有限公司

FEATURES 项目亮点

Designed in neo-classical style, it follows the ideas of "culture, sophistication, elegance, nature and privacy" to carefully create every detail.

项目以新古典主义景观风格为设计理念，遵循"文化、精致、优雅、自然、私密"的设计原则，臻于每一处点滴的创作与营造。

Site Plan 总平面图

Overview 项目概况

The project is located in Weiyang district of Yangzhou, sits on Wenchang Road, the Fuda Road in the east with the municipal government here, south of Wenchang Road opposite the boulevards Park, west of the Customs Building to Weiyang Road north to sailing Hotel business community first-line. Planning a total land area of about 50,000 m², total construction area of about 160,000 m². The project consists of the noble residential communities, leisure clubs, 5A-class financial and business office. Design using new technologies, new processes, and the introduction of "intelligent, green" elements to take full account of the interaction and coordination of the surrounding buildings and cities, using a double first floor, the double lobby, sky garden, roof garden design practices will form a three-dimensional, multi-level ecological, green, healthy space.

This project is designed to uphold the CIMC "self-reliance, to challenge the limits" spirit, with noble and exquisite landscape design of the main ideas, neo-classical landscape style design concept, follow the "culture, sophistication, elegance, nature, privacy design principles, are reaching every bit of creation and construction. Landscape design is mainly based on the neo-classical style and theme based on the analysis of the project, trying to build a new "classic elegance, sophistication and calm" commercial office and residence luxury high-end cultural landscape environment.

本项目位于江苏省扬州市维扬区东西中轴线文昌路中心位置，东至富达路与市政府一路相隔，南至文昌中路对面为街心公园，西邻海关大厦至维扬路，北至顺水大酒店商界一线。规划总用地约50 000 m²，总建筑面积约160 000 m²。该项目包括高尚居住社区、休闲会所、5A级金融商务办公楼等。设计使用了新技术、新工艺，引入了"智能化、绿色环保"元素，充分考虑了与周边建筑物的协调及城市的互动，采用双首层、双大堂、空中花园、屋顶花园等设计手法，形成立体的、多层次的生态、绿色、健康空间。

本方案设计秉持中集集团"自强不息，挑战极限"的精神，以尊贵与精致为景观设计的主要思想，以新古典主义景观风格为设计理念，遵循"文化、精致、优雅、自然、私密"的设计原则，臻于每一处点滴的创作与营造。根据对本项目的分析，景观设计主要是以新古典主义风格及主题，极力打造一个全新的"古典优雅，精致从容"的商业办公及居住的豪华高档的人文景观环境。

Commercial and Office Space 商业办公空间

Commercial office space through a full grasp of the architectural style, LOGO wall, paving materials, water features, lamp posts, sculptures, sketches, commercial umbrella chair, city landscape space is divided city markers such as extending the calm atmosphere of modern architecture the image; city commercial plaza requirements permeability to showcase their stylish elegance and atmosphere, plants to the point, the linear main, not too much, mainly native trees, rational arrangement of rare plant species to create exotic; living space combined with project planning and layout, landscape design is divided into six regions form: central garden area, sports and leisure clubs landscape area, the main entrance of the landscape area, business district and villa area to express the neo-classical style landscape.

商业办公空间通过对建筑风格的充分把握，在铺装材料、水景、灯柱、雕塑小品、商业 LOGO 墙、伞架椅、城市的景观空间划分及城市的标识物等延续大气沉稳的现代建筑形象；城市商业广场要求通透性，以展示其时尚高贵大气的一面，植物以点、线状为主，不宜过多，同时以本土树种为主，合理布置珍稀植物树种来营造异域风情；居住空间结合项目的规划布局，景观设计主要分为六大区域：中心花园区、运动休闲区、会所景观区、主入口景观区、商业区和别墅区等来表达新古典主义风情园林景观。

Residential Space 住宅空间

1. Luxurious main entrance plaza, concise flower garden Community walls, water or landscape stone leaving more space for sight of the people rather than the pace of people.

2. Large areas of carpet celebration soft lawn, dissemination of dirt and grass mixed fragrance, like a screen like the tall trees, isolated noise as much as possible to appreciate a different kind of quiet.

3. Prepare the children for the children interesting world, fully to the kids to run wild, playing, exercise opportunities, enjoy willful childhood.

4. Pavement module added natural the random planting space, and clever architectural space combined approach, so that people invariably beauty coming together to celebrate a fun-filled life.

5. The light wells introduction of fashion design concept, the integration of design and landscape design of the light wells, designed to focus on the good ventilation and natural lighting, low-carbon environmental protection concept in line with the current international advocate.

6. Designed to meet the fire road requirements mainly in order to increase the comfortable living environment and security environment of the residential area, creating an atmospheric luxury residential area.

7. A noise buffer to the immediate area of the building in a residential area, dense planting to reduce the interference of outside noise for residential areas and improve regional and quiet living environment.

1. 豪华大气的主入口广场，简明的花圃围合社区标志墙，跌水与景观置石将更多的空间留给了人们的视线而不是脚步。

2. 大片绒毯般柔软的庆典草坪，散发泥土和青草混合的芬芳，如同屏幕般的高大乔木，隔绝喧哗，尽可体味别样的静谧。

3. 专为孩子们准备的孩趣天地，给孩子们充分撒野、玩耍、锻炼的机会，享受恣意的童年时光。

4. 通过铺装模块中加入自然、随机的种植空间，以及巧妙地与建筑空间相结合等手法，使人们不约而同地因美景而聚合在一起，欢享生活的情趣。

5. 采光井引入时尚的设计理念，将采光井的设计与景观设计融为一体，设计注重良好的通风和良好的自然采光，符合当前国际提倡的低碳环保理念。

6. 设计以满足消防车道要求为主，以求提高住宅区内的居住舒适环境和安全环境，营造一个大气高档豪华的住宅区。

7. 住宅区围墙内到建筑的周边一带为噪音缓冲区，以浓密种植为主，减少外界噪音对居住区的干扰，保障区内安静的居住环境。

NEOCLASSICAL STYLE
新古典主义风格

KEY WORDS 关键词

NATURAL AND ECOLOGICAL
自然生态

LANDSCAPE ELEMENT
景观元素

LANDSCAPE NODE
景观节点

Location: Xi'an, Shaanxi
Developer: Xi'an Ronghui Properties Development Co., Ltd.
Landscape Design: Shanghai U+D & Associate
Project Area: 61,667.42 m²

项目地点：陕西省西安市
开 发 商：西安融辉房地产开发有限公司
景观设计：上海地尔景观设计有限公司
项目面积：61 667.42 m²

Xi'an Jinhui Swan Bay
西安金辉天鹅湾

FEATURES 项目亮点

The project creatively connects the neoclassic romantic emotion of meditating on the past with ancient Chinese unique architectural culture in Han dynasty; at the same time, it takes modern people's demand for comfortable life into consideration and focuses on every detailed creation.

项目创造性地将新古典主义风格的怀古浪漫情怀与古代中国特色的汉代建筑文化相联系，同时又与现代人对舒适生活的需求相组合，臻于每一处点滴的创作与营造。

Design Concept 设计理念

Following the architectural design features, the project fully displays the elegant Germany swan-castle landscape type; by using the modern language, it interprets the perfect combination of romance and elegance, deduces the low-key luxury and nobility and enjoys high-quality and comfortable life. It also creatively connects the neoclassic romantic emotion of meditating on the past with ancient Chinese unique architectural culture in Han dynasty; at the same time, it takes modern people's demand for comfortable life into consideration and focuses on every detailed creation, modern, elegant and fashionable, rather than have all the landscape elements in simple pile, stuffed the space or in messy decoration.

金辉天鹅湾遵循建筑设计特色，演绎德国天鹅堡式园林优雅景观风格，通过现代的形式语言，诠释浪漫与优雅的完美融合，演绎低调中的奢华与高贵，尊享高品质的舒适生活。项目创造性地将新古典主义风格的怀古浪漫情怀与古代中国特色的汉代建筑文化相联系，同时又与现代人对舒适生活的需求相组合，臻于每一处点滴的创作与营造，兼容典雅时尚现代，而不是将所有的景观元素进行简单的堆砌、塞满空间以及杂乱的装饰。

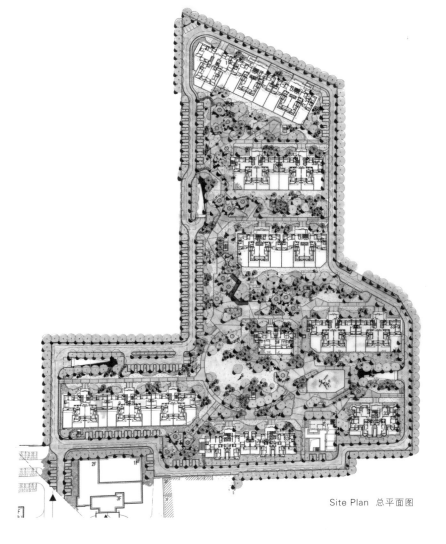

Site Plan 总平面图

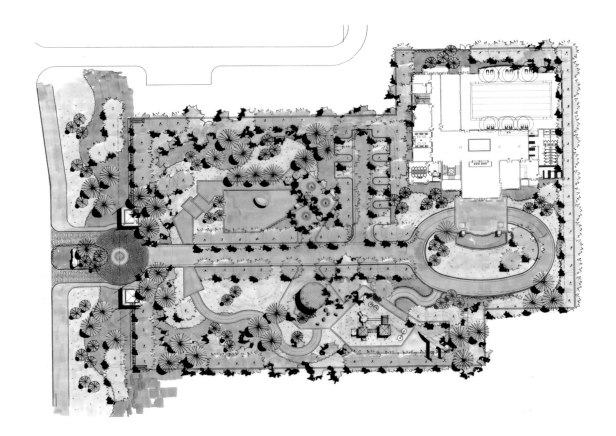

Landscape Style 景观风格

The whole design scheme of Jinhui Swan Bay goes on the premise of comfortable life, and on the basis of the neoclassical design style, it innovatively blends the inspiration of Fauvism Matisse style, using abundant landscape elements to create landscape with strong taste of art with natural and ecology throughout the planning. Water, stone, green, sculpture and rare tree species, all these elements properly shows a kind of strong vitality and bright and cheerful spring breath, forming a beautiful scenery scroll.

金辉天鹅湾整个方案构思，在舒适生活的前提下，在新古典主义设计风格的基础上，创新性地融合野兽派大师马蒂斯风格画作的灵感，用丰富的景观元素创造了一个具有艺术气息的园林，自然生态贯穿于整个规划中。水、石、绿化、雕塑、珍稀树种……所有元素恰到好处的结合，展现出一种旺盛的生命力，洋溢着春天的气息，明朗而欢快，形成一幅幅美的画卷。

NEOCLASSICAL STYLE
新古典主义风格

KEY WORDS 关键词

SUNSHINE HOME
阳光家园

LANDSCAPE CLUES
景观线索

LANDSCAPE NODE
景观节点

Location: Qingpu District, Shanghai
Landscape Design: Shanghai WEME Landscape Engineering Co., Ltd
Land Area: 1,200 m²

项目地点：上海市青浦区
景观设计：上海唯美景观工程设计有限公司
占地面积：1 200 m²

Shanghai Palaearctic Sheshan Villa
上海古北佘山别墅

FEATURES 项目亮点

The designers derive inspiration from villa buildings; take new classical musician masterpiece as design motif, and the entire layout gives different theme to different spaces of the courtyard according to fluent and converted movement composition.

项目设计从别墅建筑衍生灵感，以新古典主义音乐家的代表作为设计母题，依照流畅变换的乐章结构来谋篇布局，赋予庭院不同空间以不同的主题。

▶ Overview 项目概况

The project is located in Palaearctic Sheshan at Zhao Lane, Shanghai, on the western side of Jiasong Road, southern side of Huqingping Expressway, close to Beiganshan bird sanctuary. The villa architecture faces to the south, with the hill and the lake nearby, with complete ancillary facility, and which is of a neoclassical style.

项目位于上海赵巷古北佘山，嘉松公路以西，沪青平高速以南，紧邻北竿山鸟类保护区，别墅建筑为新古典主义风格，坐北朝南，曲水相依，配套设施齐全。

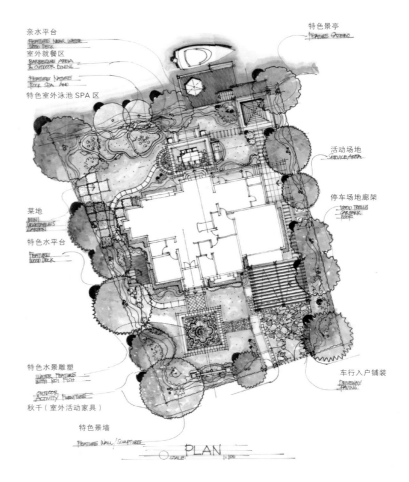

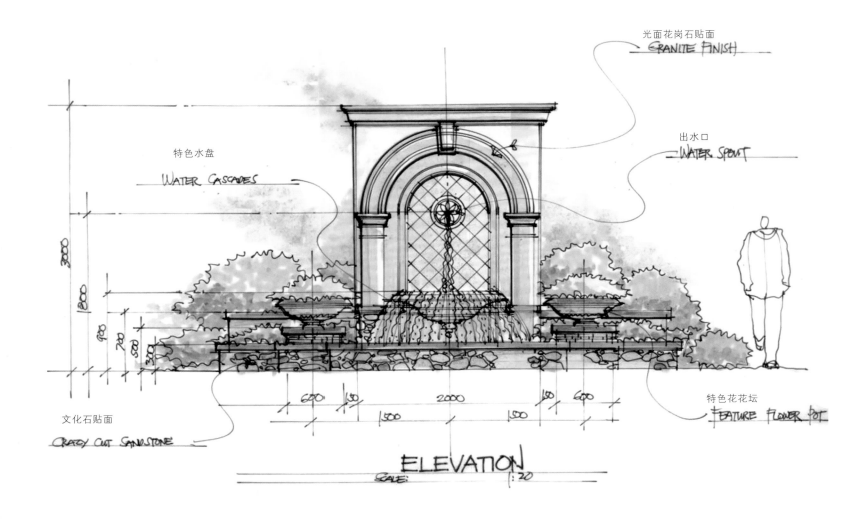

Design Concept 设计理念

The designers derive inspiration from villa architecture, take new classical musician Stravinsky's masterpiece Violin Concerto in D as design motif, and the entire layout gives different theme to different spaces of the courtyard according to fluent and converted movement composition. The design of villa garden highlights in abundant outdoor activities function, strive to create "the living room in the garden", build a dreamy home with nature, elegance, full of culture feeling.

设计灵感缘起"建筑是凝固的音乐",从别墅建筑衍生灵感,以新古典主义音乐家斯特拉文斯基的代表作D大调小提琴协奏曲为设计母题,依照流畅变换的乐章结构来谋篇布局,赋予庭院不同空间以不同的主题。别墅花园以丰富室外活动功能为设计亮点,倾力打造"花园中的起居室",营造自然、典雅、富于文化气息的户外梦想家园。

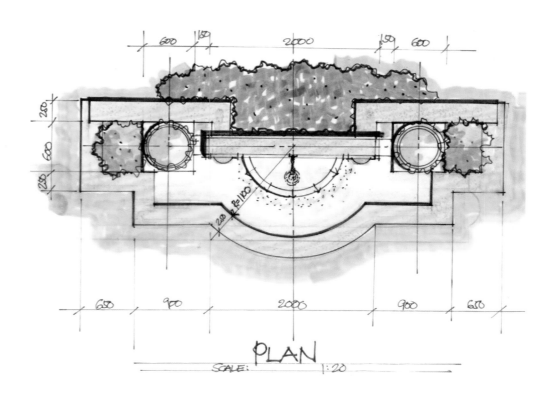

Landscape Construction 景观营造

The designers combine music and garden space via synesthesia, the entrance of the shade grassland is a semi-closed space, western water bowel as its centre waterscape highlights cheerful and lively villa entrance space. Semi-illicit and semi-closed timber frame footpath is graceful and natural as one word, three sighs style, spiral rising aria. The private and secure backyard gathers the main function of family party and social intercourse dynamic and competitive, which is like a piece of capriccio with one climax after another, flowing artistic concept and unique charm.

The designers considered the garden living room as the architectural concept to divide the space functionally; the forecourt for traffic circulation is like a hall, a group of feature wall as the opposite scenery of front gate to welcome the owner. The embellished frame connects antechamber and backyard, the material of log integrates in plant landscaping, like a sculpture made by the nature, amiable and demure, like aelegant small study. The backcourt with gathering hall, reception room, playground and vegetable garden in it, functions as owner's daily activitiy site, BBQ, dinner, rest for overlook, mini-golf, marina connected with each other but independent, satisfies family members' demand, becomes a sunshine, graceful and romantic dreamy house.

Another classic of concerto is throughout graceful chord. The designers applied musical continuity into villa landscape. Using the fireplace which is necessary in neoclassicism as dominant clue, the fireplace connect interior and exterior landscape element, and using the ubiquitous delicate waterscape as recessive clue, via penetrating and corresponding with each other the interior and exterior landscape as if fused together. People can't feel where the ending of nature is and where the beginning of art is when living in this villa.

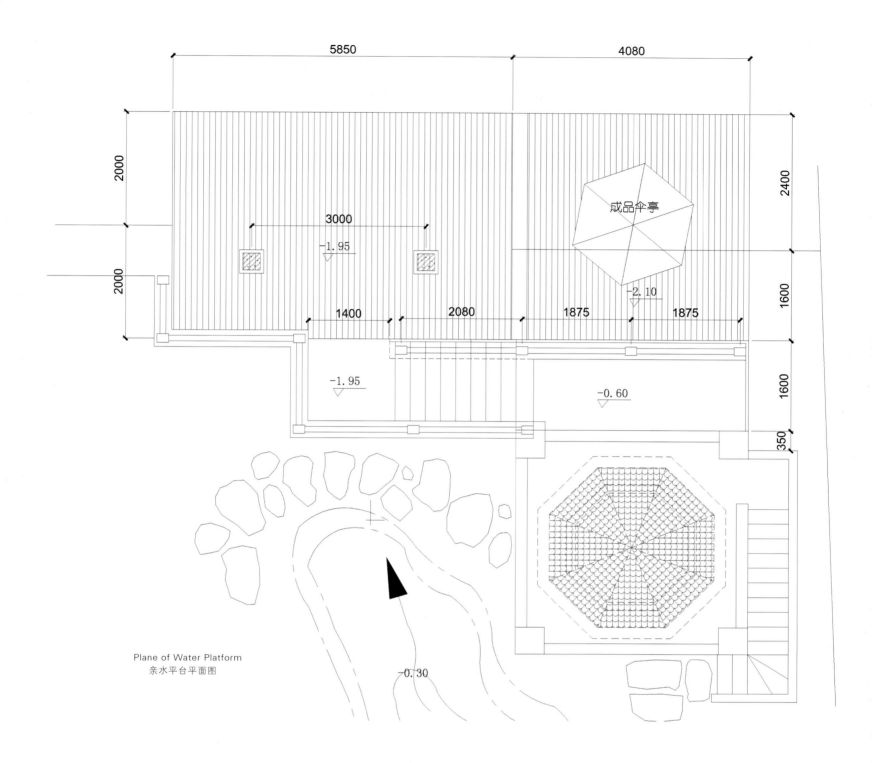

Plane of Water Platform
亲水平台平面图

　　设计师用通感和联觉把音乐和花园空间完美地结合，绿荫草地的入口开放空间，以欧式水钵作为中心水景，突出欢快活泼的别墅入口空间意向。半私密半围合的木构架步道，柔美自然，犹如一唱三叹、盘旋上升的咏叹调。后园私密安全，动静相宜，集中了家庭聚会、休闲社交的主要功能，宛如一部高潮迭起的随想曲，气韵流淌，别具风韵。

　　设计师以花园起居室的建筑概念对庭院进行功能分区，往来交通的前院好像门厅，一组景墙作为前门的对景，欢迎主人的归来。木廊架串联起前厅和后院，原木材质结合植物造景，浑然天成，亲切娴静，宛如优雅的小书房。后院集餐厅、会客室、运动场、菜园于一身，承担了供业主日常活动的主要功能，聚餐烧烤、休憩眺望、微型高尔夫运动，游船码头，不同活动空间相互联系又各自独立，满足了业主家庭成员多层次的使用需求，打造了充满阳光、美好浪漫的梦想家园。

　　协奏曲的另一个经典之处在于贯穿始终的优美和弦，设计师将这种音乐的连续性做到了别墅景观里。用新古典主义不可缺少的壁炉作为联系室内外景观元素的显性线索，以无处不在的精致水景作为隐性线索，穿插呼应，室内外的景观仿佛融为一体。人们居住其中，感觉不到自然在哪里终了，艺术从哪里开始。

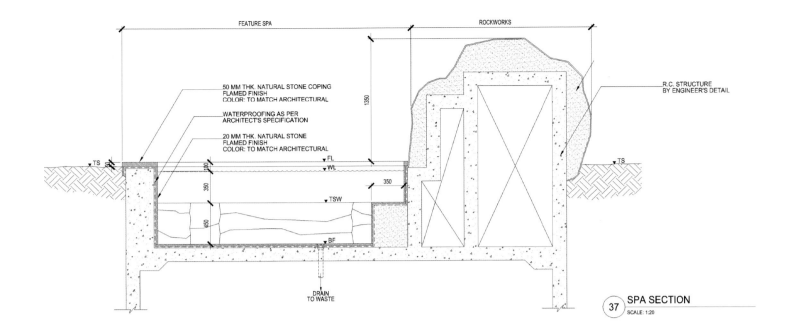

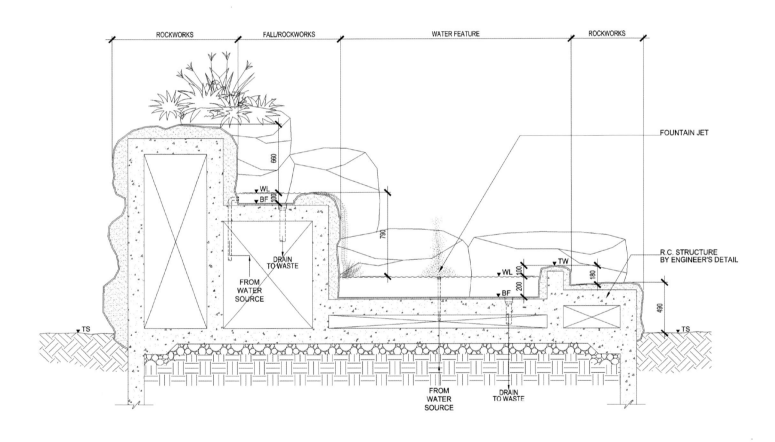

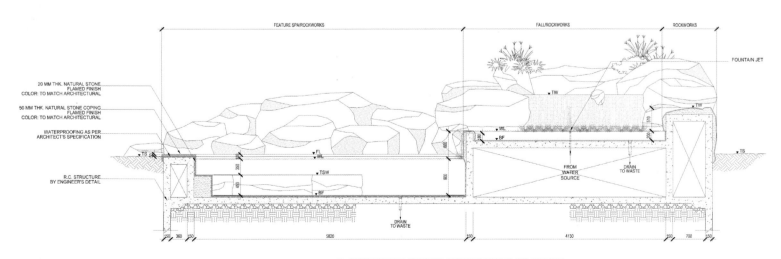

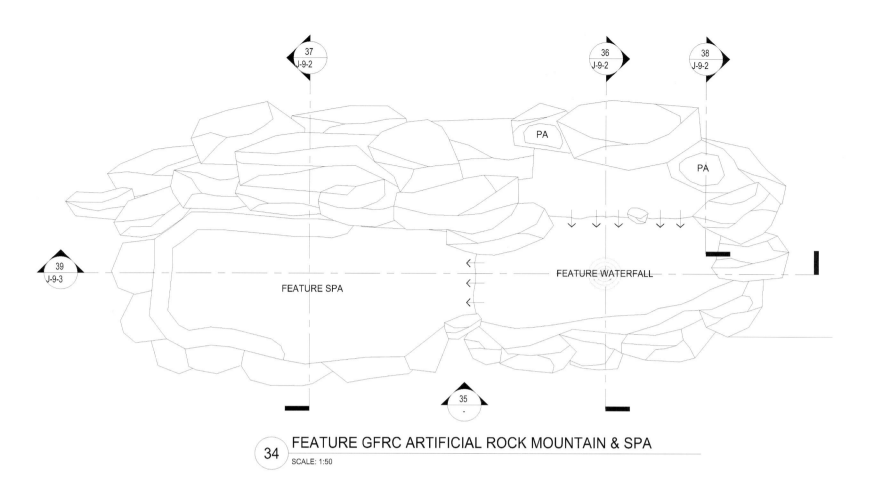

34 FEATURE GFRC ARTIFICIAL ROCK MOUNTAIN & SPA
SCALE: 1:50

NEOCLASSICAL STYLE
新古典主义风格

KEY WORDS 关键词

DECORATION ART
装饰艺术

CLASSICAL ELEMENT
古典元素

LANDSCAPE NODE
景观节点

Location: Chengdu, Sichuan
Landscape Design: Shenzhen CSC Landscape Design & Construction Co.Ltd.
Chief Designer: Li Jing
Project Director: Zhou Jiong
Solution Team: Peng Yan, Li Jia, Cao Luqian, Ming Chunhong
Construction Drawing Director: Huang Yisheng
Construction Drawing Designer: Ma Jing, Liu Qingqing, Rao Rui
Plant Designer: Luo Jingqian

项目地点：四川省成都市
景观设计：深圳市赛瑞景观工程设计有限公司
主创设计师：李婧
项目总监：邹炯
方案团队：彭妍、李佳、曹露茜、明春宏
施工图负责人：黄义盛
施工图设计：马晶、刘清清、饶睿
植物设计：罗晶倩

Zigong BRC. Gongshan NO.1 (Phase II)
自贡蓝光·贡山壹号（二期）

FEATURES 项目亮点

By means of gorgeous classical color, lofty detail carving, revered waterfront life, comfortable courtyard, the design manages to interpret the elegant, nature and romance of European Neo classicism.

该设计通过华丽的古典色调、尊贵的细节雕刻、尊崇的水岸生活、舒适的庭院空间，成功诠释了欧式新古典的尊贵和典雅、自然与浪漫。

▶ Overview 项目概况

The plot is in the south side of Gongjing New District in Zigong City, the south is near with military training base, the west is adjacent to the arterial road of Gongjing District, Changzheng Road, the north is close to judicial organs of Gongjing New District. The planning scope of Phase II is south-eastern part of the whole plot, the total land area is 55,139 m² and the landscape area is 27,474.37 m². The terrain is so various that suitable for spatial design with varied gradation.

项目地块位于四川省自贡市贡井新区南侧，南临长途客运中心，东接军区训练基地，西连贡井区主要干道长征大道，北侧为贡井新区公检法机关。本次二期规划范围位于整个地块的东南部，总用地面积为 55 139 m²，景观面积为 27 474.37 m²。地形丰富多变，竖向高差复杂，适合做出层次丰富多变的空间设计。

▶ Landscape Feature 景观特色

The project style is positioned as a lofty community of simplified European Style, showing gorgeous classical color tone, lofty detail carving and comfortable courtyard space. In terms of landscape, the high quality waterscape community is built by making full use of rolls-and-swells terrain and combining mountain and lowland. The landscape of Phase II follows the way of Phase I, fully utilizing the advantageous condition and landscape elements owned by plot, so as to make the residential environment naturalize, and to create a natural, open, comfortable living space as well as to build a residential community with unique style. The design stresses on residential environment quality, takes full advantage of surroundings and its own environment advantages, displays unique view of hillside building and delicately construct the major space node of the community and landscape line. What is more, after overall considering the residence' light, ventilate, and view etc., the land is utilized rationally, thus increasing the availability of the land.

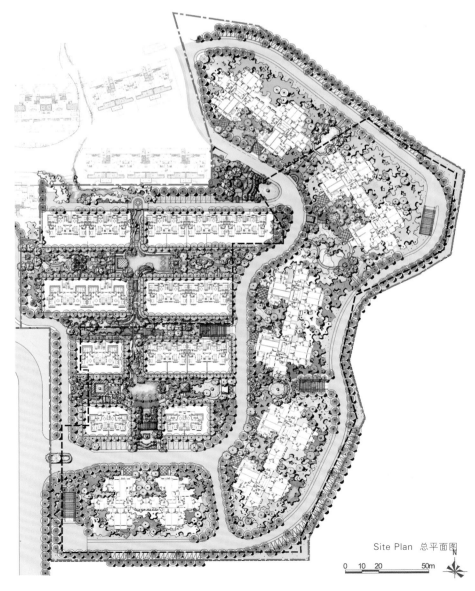

Site Plan 总平面图

项目风格定位为简欧风格高尚社区，展示华丽的古典色调、尊贵的细节雕刻，舒适的庭院空间。在景观上充分利用起伏不平的浅丘地形，结合山地与洼地，营造高品质的水景社区。项目二期景观沿用一期手法，充分利用项目用地本身赋予的有利条件及景观要素，使居住环境自然化，创造自然、开放、舒适的生活空间，营造有独特风格的住宅小区。注重居住的环境质量，充分利用周边及自身的环境优势，发挥山地建筑独特景观，精心营造小区主要空间节点及景观流线。合理节约利用土地，综合考虑住宅采光、通风、视线等综合因素，以提高土地利用效率。

Plant Setting 植物配置

Considering the indoor major landscape view, multilayer plant setting together with backyard garden builds a best viewing and entertainment space by means of leading and isolation of plant. The main purpose of high-rise landscape plant is to construct a leisure, entertainment, communication activity space and to decorate and soften building facade. The plants are also various and with distinctive characteristics.

多层区植物配置结合后花园，考虑室内主要的景观视线，通过植物进行隔离和引导，形成最佳的观景和休闲空间。设置高层景观植物的主要目的是为小区居民营造休憩、娱乐、交流的活动空间，装饰和软化建筑立面。植物的选择上也是种类繁多、特色鲜明。

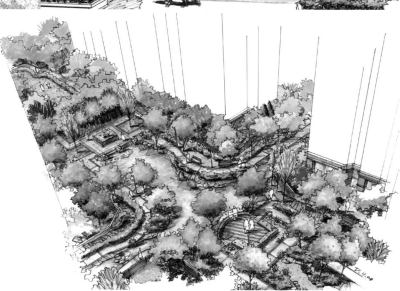

New Chinese Style
新中式风格

Oriental Culture
东方文化

Elegance
清雅含蓄

Symmetry
讲究对称

NEW CHINESE STYLE
新中式风格

KEY WORDS 关键词

SIMPLE AND ELEGANT
简约内敛

LINGNAN FEATURES
岭南特色

LANDSCAPE NODE
景观节点

Location: Guangzhou, Guangdong
Developer: Guangzhou Hehui Real Estate Co., Ltd.
Landscape Design: Guangzhou Bo'ao Landscape Design Co., Ltd.
Chief Designer: Xu Nongsi
Land Area: 32,500 m²
Floor Area: 115,700 m²
Plot Ratio: 3.00
Greening Ratio: 45.5%

项目地点：广东省广州市
开 发 商：广州市合汇房地产有限公司
景观设计：广州市柏澳景观设计有限公司
总设计师：徐农思
占地面积：32 500 m²
建筑面积：115 700 m²
容 积 率：3.00
绿 化 率：45.5%

Huidong International Garden, Guangzhou
广州汇东国际花园

FEATURES 项目亮点

Designed with the principles of plan formation and Chinese garden rules, it uses typical Chinese elements to create a modern and elegant landscape space.

设计中运用平面构成的原理和中式园林法则，以中式特有的有机图形架构为蓝本来营造现代简约的景观空间。

Overview 项目概况

The project locates in Xintang Town, Zengcheng Guangzhou. It is adjacent to the city main stem and the city greenbelt, situating in the Guangzhou Outer Ring which is in the east of Guang-Shen Expressway and the north of the Guang-Shen High-speed rail and next to National Road 107. Besides, it has convenient traffic and excellent physical environment. The project has a land area of 32,500 m² and a total greening area of 12,877 m² with a high greening ratio of 45.5%.

汇东国际花园地块位于广州增城新塘镇。该项目与城市主干道、城市规划绿地相邻，位于广州市外环线——广深高速公路以东，广深高铁以南，相邻107国道，交通便利，自然环境优美。本社区规划用地面积32 500 m²，绿地总面积12 877 m²，绿地率高达45.5%，建筑为现代简约的中式风格。

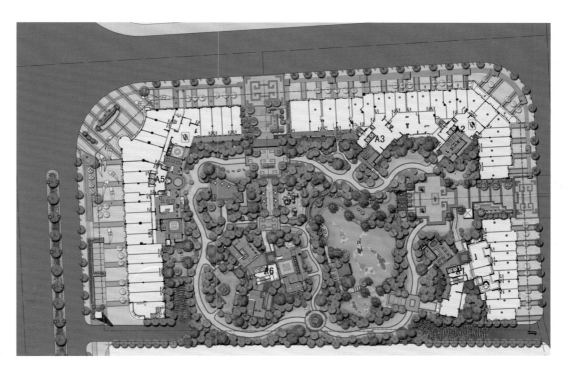

Site Plan 总平面图

Main Entrance Sectional 1
主入口区剖面图 1

Main Entrance Sectional 2
主入口区剖面图 2

Design Concept 设计理念

The rich culture heritage of Xintang leaves many sand and shells. In the ancient time, the buildings usually apply the layout of five three-courtyard quadrangles, bracket-crossing post-and-lintel structure and hard roof. The gable wall, ash plastic ridge and imbrex are the vivid portrayal of the Lingnan architecture style. However, modern Chinese architecture style is rooted in the Lingnan architectural culture and skillfully combines the eastern and western architectural culture. For the design, the project uses the principles of plan formation and Chinese garden rule, using the neat straight line, curve and fold line together to make a reasonable segmentation on the whole living area thus creating more explicit and reasonable functional space.

新塘厚重的文化底蕴积淀着无数的沙和贝，新塘古时建筑多采用五间三进院落四合院式布局，穿斗式抬梁结构，屋顶为硬山顶，镬耳山墙，灰塑脊，炉灰筒瓦，属于较具特色的岭南建筑风格之一。现代中式风格，是植根于新塘岭南建筑文化之上，并巧妙结合东西方文化的一种新思路，设计师可以在保持岭南风格的基础上以不同的形式展现。在设计中运用平面构成的原理和中式园林法则，以中式特有的有机图形架构为蓝本，通过简洁的直线、曲线和折线相结合，将整个居住区环境进行合理的分割，创造出更加明确、合理的功能空间划分。

Landscape Node 景观节点

The landscape design of Huidong International Garden strives for modern, simple, modest and luxury to reflect the Chinese modern landscape design. It carries on Chinese people's wisdom and creates boutique high-end living style of ecological high-quality residence.

The sales center lies in the northeast side, connecting National Road 107, which is one of the main display spots. It has marked stacked waterscape to call people's attention in front. Basing on the Chinese traditional afforestation technique such as view borrowing, enframed scenery and view covering, the whole community can be divided into several sequential landscape spaces to form an unbroken environment space.

Under the perfect combination of soft set and hard set, the road, trees and flower sea appropriately combine and divide every functional space. Whether the primitive entrance door or the art sculpture, sign wall and blue waterscape shows out the strong modern Chinese art breath.

汇东国际花园的景观设计，整体风格为现代简约，低调中见奢华，丰富中见稳重，充分体现整个景观现代简约的中式风情。现代中式风格情调与建筑的宁静、轻巧神韵相互融合，传承发扬中国的人居智慧，创造精品高档的生活风格，营造具有生态、人文气息的高品质心灵居所。

销售中心位于该项目东北角，与107国道相连接，是该项目最主要的对外展示景点之一。销售中心前面富有浓厚中国风情的叠级水景，形成较好的标志性作用将人们的视线引至社区。传承中国传统的造林手法，通过采用借景、框景、遮景等造景手法，做到疏可走马密不透风，将整个社区划分为若干个景观空间的过渡，并通过线条的有机分割，形成一个完整的社区环境空间。

道路、树阵、花海等运用软景和硬景的完美结合，恰如其分地围合和分割了各个功能空间，使得体块既有分隔，又相互渗透。无论是古朴古香的入口大门，还是传统中式与现代相结合的艺术雕塑、景墙、蓝色的水景等都散发出空间的强烈现代中式艺术气息。

NEW CHINESE STYLE
新中式风格

KEY WORDS 关键词

ECOLOGICAL FEATURE
生态特质

LANDSCAPE AXIS
景观轴线

LANDSCAPE NODE
景观节点

Location: Foshan, Guangdong
Developer: Foshan Hengbao Investment Co., Ltd.
Landscape Design: Guangzhou Bo'ao Landscape Design Co., Ltd.
Land Area: 66,685 m²
Floor Area: 184,805 m²
Greening Ratio: 35%

项目地点：广东省佛山市
开 发 商：佛山市恒宝投资有限公司
景观设计：广州市柏澳景观设计有限公司
占地面积：66 685 m²
建筑面积：184 805 m²
绿 地 率：35%

Royal Dragon Bay
御龙湾

FEATURES 项目亮点

Under the nature and ecology principle, the design utilizes the regional environment and arranges an optimized landscape group to build a landscape spatial series of modern, simplicity and variety.

项目在设计上遵循自然生态原则，通过对地域环境的把握和景观组团的优化配置，打造了一个现代简洁、丰富多变的景观空间序列。

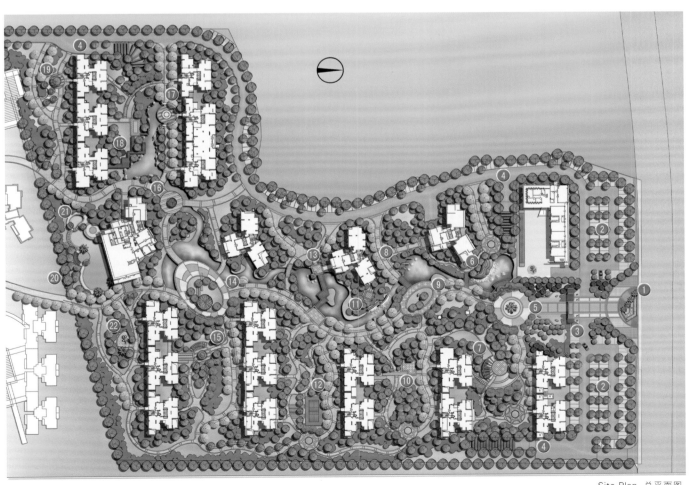

Site Plan 总平面图

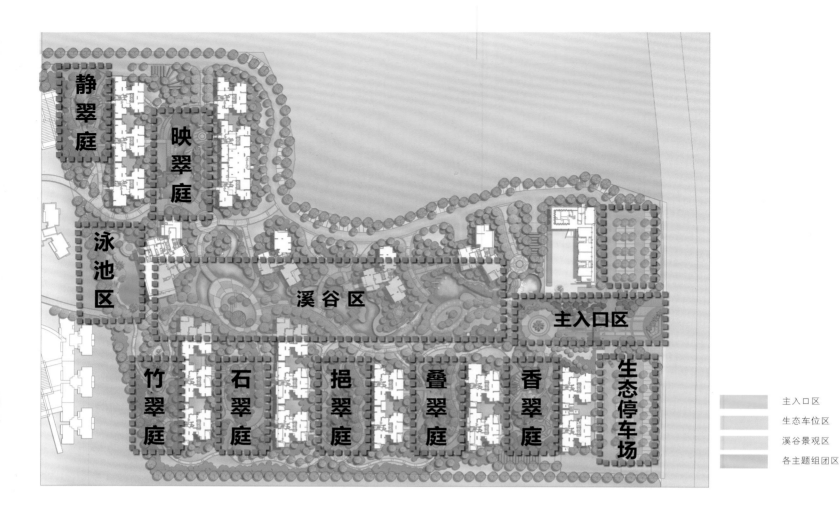

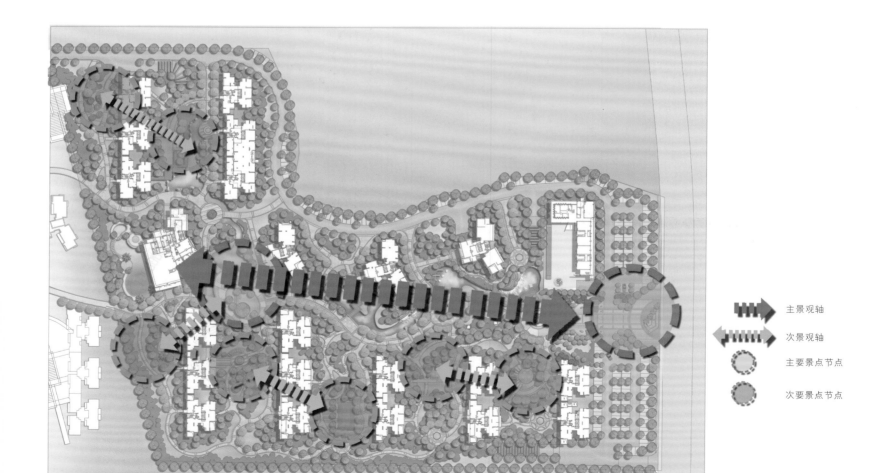

▶ Overview 项目概况

The project is located in the side of Xiqing Avenue, Sanshui District, Foshan, with a superior natural environment and convenient traffic net. According to the unique planning and design, a stream throughout the community brings vitality, interpreting the exclusive regional characteristic of Water City. Dragon Bay will be a boutique community meticulously constructed of natural and ecological features.

本案位于佛山市三水区西青大道边上，自然环境优越，交通网络四通八达。本项目规划设计别具一格，小区中一溪穿流，带来活力、生气，也充分体现了"淼城"这一特有地域属性。御龙湾将是一个被精心营造的带有自然生态特色的精品社区。

▶ Design Concept 设计理念

The design aims at reflecting the architectural design style in accordance with ideas of nature, ecology and healthy, directed by modern succinct ecology connotation to construct modern garden landscape of simplicity and grace. The integration with nature and ecology is the coral concept of the landscape design and the designers reproduce the ever-changing scenes in nature by remolding pretty nature landscape. The winding stream runs gently from south to north, acting as the background music for nature. The design has considered the balanced sharing of environmental resources for each building, guaranteeing the landscape and waterscape for each household, providing visual sense of beauty and immersive spatial effect.

设计秉承"自然、生态、健康"的思路，以体现建筑设计风格为目标，以现代简洁生态主义内涵为指导，力求打造简洁大方的现代园林景观。在项目的设计中，与自然生态的融合成为景观设计的核心思想。设计师通过对自然美景的重塑，再现自然界千变万化的景象。蜿蜒的小溪自南向北缓缓流下，长流不息的溪流声就是大自然的背景音乐……该设计不仅充分考虑到每栋建筑的均好性，做到户户有景可赏、有水可观，在视觉上给人以美感，且在空间上令人有如身临其境的感受。

Landscape Node 景观节点

The landscape design of Dragon Bay garden is summarized as "One Line and Seven Points". One Line refers to the main landscape axis—the main entrance and the stream while Seven Points is seven landscape groups. The featured pavement indicating the directions and the symbolized fallen waterscape are integrated through the main entrance landscape axis, working in harmony. The modern gate tower, sentry box and boulevards in their both sides act as the highlights of the entrance: welcoming trees, recreational plaza and featured landscape pavilion are organically organized to form varied landscape spatial series.

The featured waterscape and recreational plaza with green land in the middle are set in the entrance to keep it away from the interference of municipal roads, in terms of Fengshui. Boulevards in both sides behind the community entrance enhance the atmosphere of the entrances, and an ecological and natural landscape space is formed with tree ponds, rockery with running water, featured articles and ecological wooden bridge. The dell with natural curve is the key landscape axis of the community, which has seven different landscape nodes along both sides with different themes respectively. They are Fragrant Garden, Overlapping Garden, Emerald Garden, Rock Garden, Bamboo Garden and Serene Garden, providing visitors varied experience and sense along with every step. A great quantity of plants is adopted to soften the environment as well as different ways in dealing with the space closure.

御龙湾园林景观设计概括为"一线七点"。一线是指主入口和溪流这条主要景观轴，七点是指七个景观组团。主入口景观轴将带有导向作用的特色铺装和标志跌水水景结合，生动和谐。现代门楼岗亭和两边林荫道是入口亮点所在，迎宾树阵、休闲广场、独具特色的景观亭等景点的有机结合，形成丰富多变的景观空间序列。

主入口处的特色水景和中间设置绿岛的休闲广场，从风水的角度上讲，防止市政道路与小区入口对冲；小区门口两边特色林荫大道，增加入口气氛。现代造型的树池、假山叠水、特色小品、生态木桥组成一个生态、自然的休闲景观空间。走势自然弯曲的溪谷，是小区的重点景观轴线。溪谷两边顺势设置了七个不同的景观节点，且每个点各有主题，分别为：香翠庭、叠翠庭、挹翠庭、石翠庭、竹翠庭、映翠庭、静翠庭。还利用了大量植物软化环境，且围合空间时运用了有收有放、欲扬先抑的手法，带给参观者步移景迁的参观感受。

NEW CHINESE STYLE
新中式风格

KEY WORDS 关键词

TRADITIONAL GARDEN
传统园林

LANDSCAPE LAYOUT
景观格局

LANDSCAPE NODE
景观节点

Location: Chaoyang District, Beijing
Developer: Poly Real Estate Development (Beijing) Co., Ltd.
Landscape design: L&A Design Group
Landscape area: 34,000 m²

项目地点：北京市朝阳区
开 发 商：保利（北京）房地产开发有限公司
景观设计：奥雅设计集团
景观面积：34 000 m²

Landscape Design of Beijing Poly Oriental Mansion
北京保利东郡项目景观设计

FEATURES 项目亮点

The design emphasizes the purity of spirit and the sublimation of mentality. In terms of environment, the design does not only meet the need of aesthetic appreciation, but it also represents people's advanced spiritual demands to approach a state of 'capacious, lively and calm'.

项目强调社区的仪式感、品质感和尊贵感，在环境营造上不单单满足人的观赏性，更多的是表达人更高层次的精神诉求，使环境和心灵都达到"空、灵、静"的唯美境界。

Overview 项目概况

Located on the East 4th Ring Road, at the intersection of the high-end residential section - Chaoyang Park section and the New CBD, Poly Oriental Mansion is only 1,000 m from Chaoyang Park - the largest central park in Asia. Surrounded by Young Pioneer Park, Chaoyang Park, Lake Union Park and three new parks in the future CBD, it enjoys a great park view. In addition, as the first high-end residential project after the eastern extension of the CBD, it also enjoys the mature facilities around.

保利东郡位于北京市朝阳区东四环，高端人居板块朝阳公园板块与新CBD双核交汇处，距离亚洲最大城市中央公园朝阳公园仅1 km，被红领巾公园、朝阳公园、团结湖公园以及未来CBD区域内即将新建的三座公园六园环抱，尽享园区宜居景致；同时作为CBD东扩后区域内第一个入市的高端住宅项目，尽享周边繁华资源。

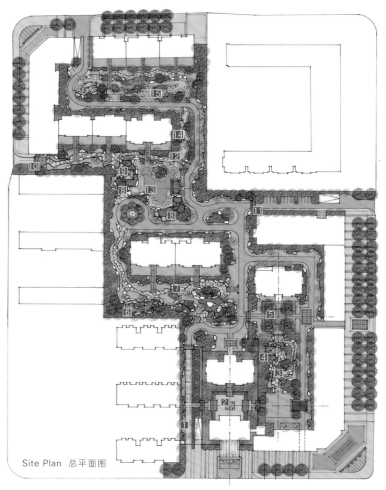

Site Plan 总平面图

Design Concept 设计理念

The project explores the modern reflection of traditional artistic conception through a new design philosophy. According to the history of the site and Poly's rich multiculture, L&A creates the landscape theme to highlight the exclusiveness and uniqueness of the design. Besides, the design attempts to make a special breakthrough in landscape concept, which emphasizes the purity of spirit and the sublimation of mentality. In terms of environment, the design does not only meet the need of aesthetic appreciation, but it also represents people's advanced spiritual demands to approach a state of 'capacious, lively and calm' which reflects the cultural connotation of Chinese people. It also emphasizes the feeling of ceremony, quality and nobility in the community.

　　该项目以全新的设计理念来探索传统意境的现代演绎，从地域历史和保利集团自身丰富的多元文化中挖掘题材，打造设计的独创性和唯一性，并力求在景观概念上有新的突破，着眼于心灵净化、思想升华，同时强调了社区的仪式感、品质感和尊贵感。在环境营造上不单单满足人的观赏性，更多的是表达人更高层次的精神诉求，使环境和心灵都达到"空、灵、静"的唯美境界。"空、灵、静"表现了一种中国人特有的文化内涵，是儒学、理学、禅学、哲学中的一种伦理、秩序和逻辑。

▶ Landscape Line 景观流线

In terms of landscape design, front courtyard and back courtyard in traditional landscape design are set. The front courtyard has shown ethics, order and logic, whilst the back courtyard takes advantage of the terrain to be surrounded by mountains and waters. Thus people will feel like living in a peaceful valley. In combination with Chinese traditional Confucianism, philosophy and Zen Buddhism, it has created a quiet environment in the hustling and bustling city.

整个景观格局取自传统园林的前庭后院布局，前庭凸显伦理、秩序和逻辑，后院以地势围合形成以山龙之稳健与水龙之灵动的双龙戏珠的山水之势，使人身在园内有幽居山谷的感觉。在人文上结合中国传统儒学、哲学和禅学，营造大隐隐于市的祥和意境。

NEW CHINESE STYLE
新中式风格

KEY WORDS 关键词

ECO COMMUNITY
原生态住区

SLOPES
坡地特色

LANDSCAPE NODE
景观节点

Location: Oriental Garden, OCT, Nanshan District, Shenzhen
Developer: Shenzhen OCT Real Estate Co., Ltd.
Architectural Design: Aodi International Architectural Design Co., Ltd.
Designers: Liang Wenjie, Pei Xiemin, Zhou Jingqiang
Lang Area: Area P1 12, 365.20 m²
　　　　　Area M3 6, 339.40 m²
Floor Area: Area P1 9, 487.04 m²
　　　　　 Area M3 4, 827.39 m²

项目地点：深圳市南山区华侨城东方花园
开 发 商：深圳华侨城房地产有限公司
建筑设计：傲地国际建筑设计有限公司
设计人员：梁文杰、裴协民、周敬强
占地面积：P1区 12 365.20 m²
　　　　　M3区 6 339.40 m²
建筑面积：P1区 9 487.04 m²
　　　　　M3区 4 827.39 m²

Zone P1+M3 of OCT East, Shenzhen
深圳东部华侨城P1+M3区

FEATURES 项目亮点

The landscape design takes advantage of its coastal location and creates a beautiful environment of modern Chinese style.

景观设计既充分地利用了滨海环境的特色，又很好地把握了现代中式风格的基本格调。

▶ Overview 项目概况

Covers an area of 200,000 m², the project is one of the pioneer seaside villa areas in Shenzhen. Located between the two theme parks, the World of the Window and folk culture village, it is at the south to Shennan Avenue and north to Shenzhen OCT Wetland Park. It is a rare ecological residential area that surrounded by numerous parks. Area P1 and Area M3 are the newly-built low-density residences in Oriental Garden.

该项目占地约200 000 m²，是深圳最早的滨海别墅区之一，位于深圳世界之窗和民俗文化村两座主题公园之间，北临深南大道，南面是深圳华侨城湿地公园，是深圳罕有的环抱于多座公园之中的原生态住宅区。P1区和M3区则属于东方花园内新建的低密度住宅。

▶ Landscape Node 景观节点

Based on the primary slope, P1 makes the best of beautiful natural scenery of OCT Wetland Park and Happy Coast to ensure every house may enjoy both the internal environment and the external landscape. Four semi-detached houses are set in M3, in which the main building has three floors. Each house connects with the path in the garden directly. In accordance with the current condition, it sets garage in the semi-open basement that provides a larger space for the private garden.

Site Plan 总平面图

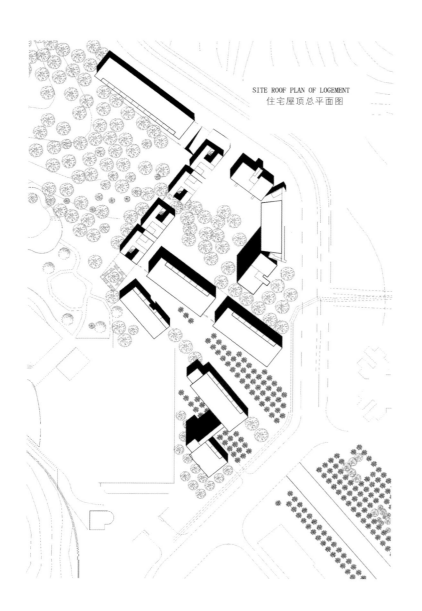

SITE ROOF PLAN OF LOGEMENT
住宅屋顶总平面图

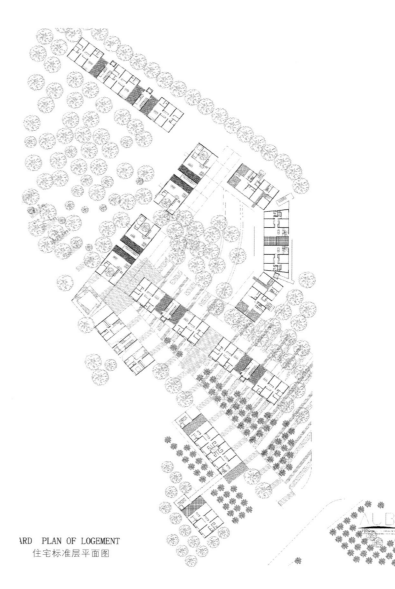

ARD PLAN OF LOGEMENT
住宅标准层平面图

Buildings in P1 are not the simple overlay of house types, but the sculptures that create elaborately. Starting from shape "L", every element inset with each according to rigor propotional relationship. Construction should take the interactive relationship between woods, lake, sunshine and terrain into account is the design principle. Each house should have a large outdoor terrace garden and private garden.

In M3, there are semi-detached villas, which highlight the relationship between the construction and the courtyard. While ensuring the privacy, it obtains sharing landscape resources with the surrounding environment. Large glass window and the roof garden enable the construction to borrow the nice scenery from the surrounding ambiance. Each house has its private garden and detached garage. Stone, exposed concrete, wood and steel are the main building materials.

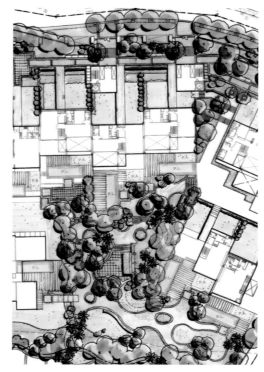
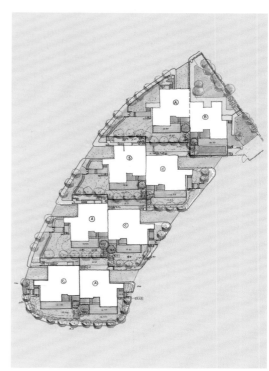

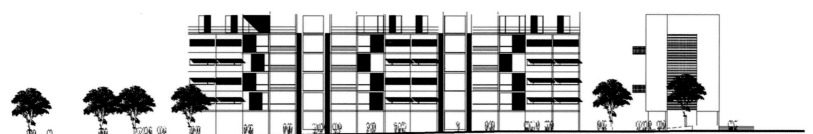

Elevation 1 立面图 1

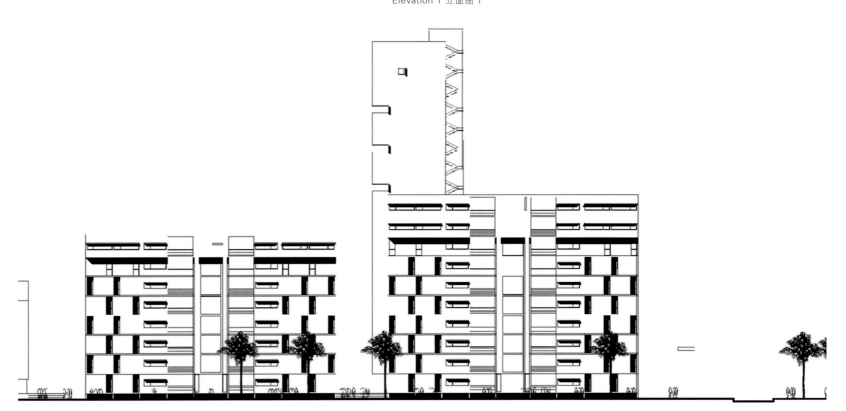

Elevation 2 立面图 2

P1区结合基地原有坡地特色，充分利用华侨城湿地公园和欢乐海岸优美自然景色，确保所有住户都拥有分享园内优美环境和观赏华侨城外部景观的权利。M3区规划设计为四栋双拼别墅，建筑主体为三层，每户都与园区道路直接相连。利用现状地形，将车库置于半地下室，充分释放地面空间，为营造私家花园创造了更大的空间。

P1区的建筑体不是户型的简单叠加，更像是经过精心打造的雕塑，设计从"L"形开始，每个细胞彼此镶嵌，按照严密的比例关系协调在一起。创作原则是建筑必须考虑与林、湖、阳光、台地互动的关系，每户都有大的室外露台花园、自己的私家花园。

M3区建筑设计为双拼别墅，注重建筑与庭院的关系。在解决良好的私密性的同时与周围环境取得景观资源的共享。大面积玻璃窗及屋顶花园平台的应用使建筑得以借用周围上佳景色。每户都有自己的私家花园和独立车库，建筑材料以石材、清水混凝土、木材、钢为主，造型稳重，色彩丰富。

NEW CHINESE STYLE
新中式风格

KEY WORDS 关键词

DELICATE STRUCTURE
结构精巧

ENCLOSED PLANTATION
植物围合

LANDSCAPE NODE
景观节点

Location: Foshan, Guangdong
Developer: Foshan Hefeng Yingyuan Garden Real Estate Co., Ltd.
Landscape Design: Guangzhou Bonjing Landscape Design Co., Ltd.
Land Area: 65,464 m²
Floor Area: 118,111 m²
Plot Ratio: 1.2
Green Ratio: 35%

项目地点：广东省佛山市
开 发 商：佛山和丰颖苑房地产有限公司
景观设计：广州邦景园林绿化设计有限公司
占地面积：65 464 m²
建筑面积：118 111 m²
容 积 率：1.2
绿 化 率：35%

Foshan Hefeng Yingyuan Garden
佛山和丰颖苑

FEATURES 项目亮点

The overall design adopts a key note of "combining the Chinese and Western elements, taking new ideas from the past", using the reference of garden construction technique in Chinese classical landscape architecture and impressing it with the western modern landscape language.

采用了"中西合璧，古韵新做"的理念，借鉴了中国古典园林的空间营造手法，并以西方现代的景观语言加以诠释。

Site Plan 总平面图

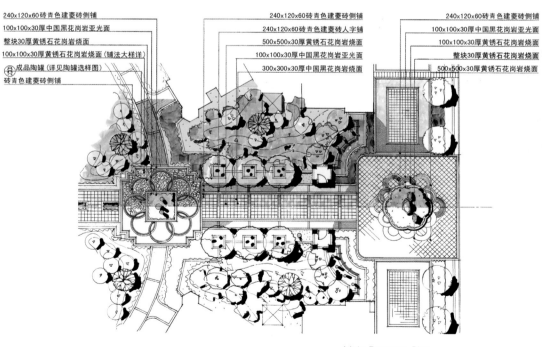

Overview 项目概况

Hefeng Yingyuan Garden is located near the Xianhu Lake of Xianhu Tourism Resort District, Nanhai district, Foshan city. It covers an area of about 65,464 m², building area of about 118,111 m² and the overall greening rate is 35%, plot rate is 1.2%. There are about 900 households. These houses vary from 36 m² a room to 180 m² four-room suite, and a small amount of 260 m² multiple units. The project is divided into three blocks. There are Xianhu Hotel, Wuji Well-being Park in the east, Guidan Road in the south, villa communication in the west, Xianhu Tourist Resort and Xianhu Bay Plaza in the north. This residential garden sits north and faces south. According to landform, the garden adopts half enclosing layout and arrangement type, giving priority to north-south apartment layout and a little east-west households as well. The garden has a north-south axis in the middle block and the other axis along Guidan Road, paralleling through the drainage net and stream landscape to connect these three blocks into an organic entirety.

和丰颖苑位于广东省佛山市南海区丹灶镇仙湖旅游度假区的仙湖之畔，占地面积约 65 464 m²，建筑面积约 118 111 m²，规划的住宅小区总户数约 900 户。户型面积区间基本由 36 m² 的一房至 180 m² 的四房组成，也有少量约 260 m² 复式单位。项目规划用地共分为三大地块，东面为仙湖酒店、无极养生园公园，南面为桂丹路，西面为别墅小区，北面为仙湖旅游度假区及仙湖湾广场。小区采用南北朝向，根据地块形态半围合布局及排列，以南北向户型为主，部分为东西向。小区内沿中间地块设南北轴线，另一条轴线沿着桂丹路平行通过水系及溪流景观延伸，使三大地块有机联系形成一个整体。

Main Entrance Plan
主入口区平面图

186

▶ Design Concept 设计理念

The overall design adopts a key note of "combining the Chinese and Western elements, taking new ideas from the past", using the reference of garden construction technique in Chinese classical landscape architecture and impressing it with the western modern landscape language. Leaded by the natural ecological principle, the green area and personality-diverse landscape space are added as much as possible to make the building group and plants symbiotic and harmony and provide the owners abundant landscape view and physically and mentally relaxed space.

设计采用了"中西合璧，古韵新做"的理念，借鉴了中国古典园林的空间营造手法，并以西方现代的景观语言加以诠释。项目以自然生态原则为主导，尽可能增加绿化面积和个性多样化的景观空间，使得建筑群体与花草林木共生共融，为业主们提供丰富的景观视野和放松身心的社区空间。

▶ Landscape Layout 景观布局

At the same time, it uses the landscape techniques such as changing views in every step, winding path leading to a secluded spot and plant enclosing to build a rich landscape sequence space. Through multilayer terrain of progressive sense of space, a closed or open, public or private, raised or sinking, connect fully or concealed space is built. The stillness of the space combines the scene and the flow of time and space affect visitors' pace. Besides, it uses plants, water, pavement, sketch, etc to create space, and through the funny linked path to provide the resident an activity place for rest, activities, fitness and so, striving to achieve the artistic conception of garden.

The landscape layout is subtle with level-heights at random and winding roads. It sets views in details, tight and not forced, quiet and interesting. The green trees make a pleasant shade and flowers blooms like a piece of brocade, water runs in circle, the flower and willow flourish along the shore.

At the entrance, the highly dynamic waterscape design interprets the classic luxury charm. Every active drop of water adds the quiet community more vitality. There is also a landscape lake, fine willow jiggling at the shore with floating fragrance. The lake sparkles in the sunlight and the fountain in the middle of the lake rushes up straightly, ably blending with gallery, pavilion and hurdles near the lake. The whole landscape design has reached the state of "neatly layout, artful structure, decorative beauty, refine design and deep cultural connotation "

设计运用了步移景异、曲径通幽、植物围合等造园手法，营造出围合的或开敞的、公共的或私密的、凸起的或下沉的、通透的或隐蔽的多样空间。利用植物、水景、铺装、小品等营造空间，并通过有趣味性的路径相连，为居民提供休息、活动、健身等不同功能的活动场所，力求达到"庭院深深深几许"的意境，方寸间自成天地，精细自然之处却包含无限深邃。

　　社区景观布局精妙，层次高低错落，道路回环，空处有景，疏处不虚，大中阔景，小中致景，密而不逼，静中有趣，幽而有芳。加上丰富的绿化种植，从而形成绿树成荫、繁花似锦、曲水回环、花堤柳岸等具有特色的景观韵味。

　　入口处，以极富动态的水景设计诠释了古典的奢华魅力，以其灵动的跌水为宁静的社区带来汩汩生机。进入园内，会被满庭青翠、错落有致的景色所吸引。在阳光雨露之中，花草吐露芬芳，显得更加多姿多彩，生机盎然。园内设计了景观湖，湖岸细柳轻摇、暗香浮动，湖面波光粼粼、疏影横斜，湖中喷泉直涌，结合湖边廊、亭、栏，巧妙地构成一体。整个景观设计既有大形式上的统一，也有细节上的精雕细琢，达到了"布局之工，结构之巧，装饰之美，营造之精，文化内涵之深"的境界。

NEW CHINESE STYLE
新中式风格

KEY WORDS 关键词

NATURE & ECOLOGY
自然生态

ARTISTIC STYLE
艺术风格

LANDSCAPE NODE
景观节点

Location: Guilin, Guangxi
Landscape Design: Donglee Landscape Group
Designer: Cao Xing

项目地点：广西壮族自治区桂林市
景观设计：当代东篱建筑景观机构
设 计 师：曹星

Zhangtai Wisdom City, Guilin
桂林彰泰睿城

FEATURES 项目亮点

The landscape design with oriental health garden as the theme, adopts the integration of Chinese garden and Japanese garden.

景观设计以东方养生园林为主题，表现为中国园林和日式园林结合的形式。

Overview 项目概况

In responds to the architectural planning and design, the landscape design of Wisdom City takes "one ring, two axes, three levels of water system and four groups" as its basic structure. One ring refers to the main lane in inner ring shape in the community. Two axes are landscape axes in north-south direction and west-east direction. Three levels of water system comprise the landscape swimming pool in central zone, hydrophilic zone in the inner side of main lane, and streams surrounding the architecture groups. Four groups refer to four zones divided by roads, with green land in inner side as leisure activity zone equipped with recreational and health facilities.

桂林睿城景观设计以建筑规划设计为基础，以"一环，二轴，三层水系，四大组团"为基本结构。一环是小区内环形主车道，是区内主要的道路。二轴是东西景观轴和南北景观轴。三层水系包括中心区景观泳池，主车道内侧亲水休闲带和外围建筑边的溪流。四大组团是以道路划分的四个区域，以内向绿地为休闲活动区，其间安排各种休闲健身设施。

Master Landscape Plan 景观总平面图

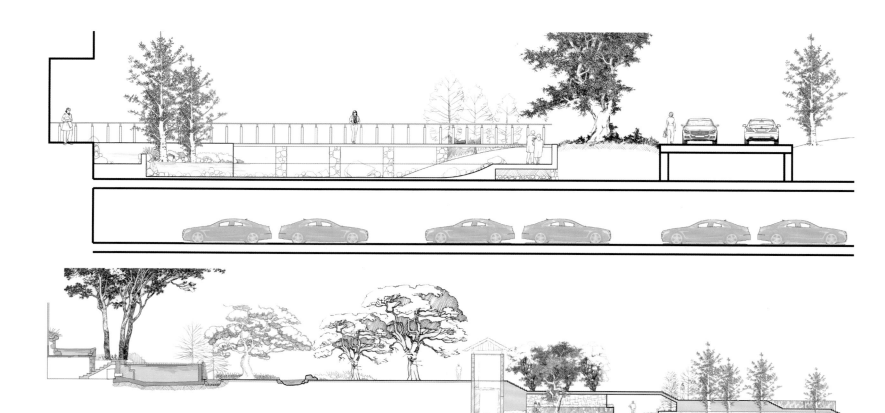

Design Concept 设计主题

The landscape design with oriental health garden as the theme, adopts the integration of Chinese garden and Japanese garden. The design of Japanese garden is in natural style deeply affected by Chinese garden style especially landscape gardens in Tang and Song Dynasty, consequently similar to Chinese gardens. However, it is integrated with Japanese natural conditions and cultural background, establishing its own system with unique style. Japanese garden pursues on liberal style with poetic and philosophic connotation. The project design is a integration of Chinese and Japanese garden, with tree greening as the base, building a favored garden landscape featuring dynamic waterscape and health-theme garden.

睿城景观设计以东方养生园林为主题，表现为中国园林和日式园林结合的形式。日本深受中国园林尤其是唐宋山水园林的影响，因而一直保持着与中国园林相近的自然式风格。但结合日本的自然条件和文化背景，形成了它独特的风格并自成体系。日本园林讲究造园意匠，极富诗意和哲学意味，形成了"写意"的艺术风格。设计中融合了中日园林的精华，以大树绿化为基础，结合动态水景和养生主题园中园，打造宜人的园林景观。

Landscape Node 景观节点

The design of the entrance is the key step in the project. The entrance in south is mainly for vehicles, with a cherry avenue. The entrance in west is mainly for pedestrians, connecting to external Commercial Street, and right facing Lidong Park. As an open design, a row of trees and a iconic sculpture is the proposing landscape while drop waterscape and terraces are positioned in the medium part to deal with the problem of height difference, forming three-dimensional traffic upper and lower.

Barrier-free access in the lower level is equipped with residential elevator for uses, showing the theme of human orientation.

The central zone has heated swimming pool club and landscape swimming pool outdoor. Swimming pool outdoor acts like a natural hot spring in mountain forest, integrated with the terrain forming a waterfall and a source of the streams. The swimming pool zone is divided into children pools and adult pools. The club building is set off by the rocks and trees, like growing in nature. Large-sale plaza is set in east die of the club for inhabitants' activities.

A stream connects the household terraces outside the buildings, fully demonstrating the design concept of hydrophilicity, nature and ecology. The simple and smooth lines and running water system enhance the natural and ecological touch in the park. Places for inhabitants' social communication are positioned around the stream, with the scattered arrangement of waterfront plaza, water platform, boardwalk, water streams, fountains, and springs offering experience and fun of waterfront dwelling.

Six garden parks are designed in four groups, sharing the same great theme of Health Park while boasting their own characteristics respectively, showing oriental garden style. They are Taiji Garden, Puti Garden, Qin Garden, Qi Garden, Xijian Pool (Shu Garden), Huazhongyou (Hua Garden). The design of garden in park features pure, natural and exquisite, leading a residential park from functional design to artistic design.

入口的设计是重点。南入口是主要车行入口，两边种植樱花形成樱花大道。西入口是主要的人行入口，入口与外围商业街相连，正对滴东公园。因此入口采用开敞式设计，用一排大树和一个标志性雕塑为前段景观，中段采用跌级水景和台地设计处理高差问题，形成上下层的立体交通。下层是无障碍通道，在尽头设计入户电梯，体现"以人为本"的主题，方便住户使用。

中心区是恒温泳池会所和户外景观泳池。户外泳池设计成山林中的自然温泉形式，结合地形地势，泳池落水形成一条山林瀑布，作为溪流的源头。泳池区分为儿童池和成人泳池。大量的山石和树木的掩映，使会所建筑仿佛从山林中自然生长出来，别具风貌。会所东面设计有大型活动广场，供住户集中活动。

外围建筑以一条溪流连接各个入户平台，充分展现了"亲水"及"自然生态"的景观设计概念。流畅简洁的水岸线与灵动水系无形中提升了园区的自然生态气息。围绕溪流周边设置了住户交流的场所，并通过临水广场、亲水平台、木栈道、戏水溪涧、喷泉、涌泉的穿插铺排，让居者体味临水而居的悠然与乐趣。

四大组团内设计了六个特色园中园，各自统一在养生园林的大主题下，又独立成园，体现东方园林的神韵。共分别是太极园、菩提园、琴园、棋园、洗砚池（书园）、画中游（画园），以清纯、自然、小巧为特色的园中园设计，把居住区园林从功能设计提升到艺术设计的高度。

NEW CHINESE STYLE
新中式风格

KEY WORDS 关键词

LANDSCAPE ELEMENT
景观元素

GROUP SPACE
组团空间

LANDSCAPE NODE
景观节点

Location: Yili, Xinjiang
Developer: Xinjiang Yili Renhe Real Estate Development (Group) Co., Ltd.
Landscape Design: Shenzhen Joco Landscape Design Co., Ltd.
Area: 91,340 m²

项目地点：新疆伊犁
开 发 商：新疆伊犁仁和房地产开发（集团）有限责任公司
景观设计：深圳市筑奥园林景观设计有限公司
项目面积：91 340 m²

Yili Renhe - Ningyuan County
伊犁仁和·宁远郡

FEATURES 项目亮点

The design aims to create the most comfortable home in beautiful Xinjiang and let one feel the local customs and nature rhyme in Yili. The construction focuses on peace and upholds essence of exquisite in Chinese culture.

设计旨在创造边疆最适宜的家，让居住者尽情享受伊犁的风情与中式的文化气息，通过对本土文化的千锤百炼，对古典文化的精挑细选，凝练出极具特色的设计符号。

▶ Overview 项目概况

The project is located in the Border Economic Cooperative Zone of Yining City in Yili Kazak Autonomous Prefecture, Xinjiang which enjoys a convenient traffic, smooth terrain, fresh air and beautiful scenery. It is the strategic region for economic development in Ili River Basin which enjoys favorable location, policy and environment.

伊犁仁和·宁远郡位于有"塞外江南"之称的伊犁哈萨克自治州伊宁边境经济合作区。北邻218国道，南沿美丽的伊犁河畔，交通便捷、地势平坦、空气清新、风景优美，地处伊犁河流域发展经济的战略区域，具有良好的区位、政策、环境优势。

▶ Inspiration 设计灵感

With the goal to create beautiful and comfortable homes on the boarder, enable residents to enjoy typical Yili style and Chinese culture, it combines local culture with traditional Chinese culture to create unique design elements. The inspiration comes from the culture of "Harmony" (和) in "The Analects of Confucius" and "The Book of Songs"(two famous collections of ancient Chinese poems and classics). The spaces symbolize the so-called "five cardinal virtues" in traditional Chinese culture — benevolence, righteousness, manners, wisdom and faithfulness, which inspire people to achieve lifelong laughter and happiness.

The design respects the historical context, combining local conditions with modern living ideas. Rustic Yili landscape is the main theme. Together with the oriental flavor and the traditional Chinese culture, it restores an elegant and dignified modern Chinese "courtyard" for views, living and enjoyment. It will be an ideal home for the Confucianism followers.

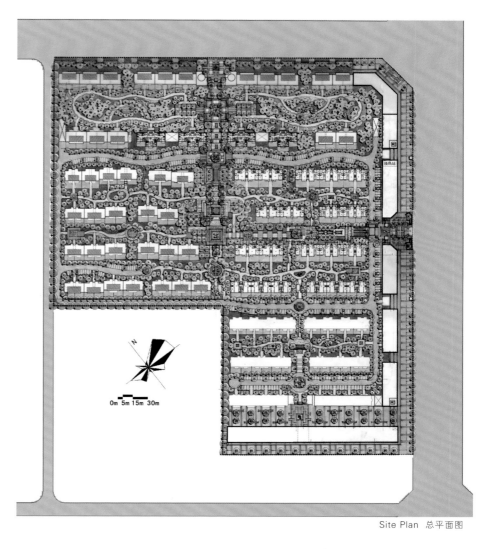

Site Plan 总平面图

　　设计旨在创造美丽边疆最适宜的家,让居住者尽情享受伊犁的风情与中式的文化气息,通过对本土文化的千锤百炼,对古典文化的精挑细选,凝练出极具特色的设计符号。设计灵感源自《论语》与《诗经》,深入中国文化之精髓——"和",传习中国人文血脉中"精雕细刻"的精华。几个递进的空间分别代表"仁、义、礼、智、信"等深层含义,启示人们获得终生的快乐与幸福。

　　项目传承历史文脉,汲取本地的气候、人文风貌精髓,引入现代的人居理念,以温淳质朴的伊犁自然风情作主轴,以极具包容性的东方意境及中式古韵为脉络,再造现代中式"前堂后院"典雅、尊贵的景观,可赏、可游、可居、可品,成功演绎了儒家文人的理想家居环境。

▶ Landscape Space　景观空间

Landscape walls serve as the partitions between courtyards. Modern landscape skills are adopted to create modern oriental elements. Red pavements, embedded reliefs, gray walls, cascade walls, etc., all show classical Chinese style. The facade is mainly made from red bricks, black tiles, concrete, glass and steel. Tiles and bricks as the important elements for local dwellings, are largely used to shape the local flavor of the project.

Courtyard, as "a treasured place" in Chinese "Feng Shui", is the typical Chinese residence style that conforms the national character of the Chinese people. In this project, courtyard landscape is well designed in natural style to increase daily communications between family members. Different landscape spaces compose different courtyards, and the landscapes along central axis provide spaces for meeting and activities. Within the building groups, natural green landscapes follow the topography. In addition with the flowers and grass, it provides the residents with peaceful and romantic courtyard spaces.

Lanes go forward one by one to connect the courtyards from the "enhance" to the "hall", embodying the idea of "benevolence and harmony". Landscape changes as one walk through the courtyards, providing the experience of great Yili. Lanes lead people back home, along which, flowers, trees and grass change in different seasons to present different views.

景观设计用景墙分隔开不同的院子，利用框景与对景的手法，用不同的形体，体现景观现代东方元素，大面积的红砖铺装、嵌入浮雕、灰色墙体、瓦片堆叠的跌水墙，带有浓郁的中式情怀。红砖、青砖、混凝土、玻璃、钢成为主要的立面材料。砖瓦作为大量普通居民的重要元素，以它为主基调，赋予园区鲜明的地方特色。

庭院，是中国风水上所追求的"藏风聚气"之地，它与中国人内敛含蓄的民族性格相协调，是风格鲜明的中式居所。此项目内敛型的院落让每个庭院更自然，家庭成员之间更有亲切感，院子的景观规划让家的氛围更浓。设计用不同的景观空间组成不同的院落，中轴景观以展示型为主赋予集散与活动空间，组团空间以自然绿化为主，自然的地势，各种流线形的花草，给予居住者安静浪漫的庭院。

街巷由一个个递进式庭院组成。从开始的"门"主轴的展示区到"堂"中轴的参与式，以门为仁、庭为和的框架，表现了"仁和"的意涵，体现了仁义之士尊尚礼节，门庭华丽，行走间悠然自得，移步异景，景景雅静，凸显塞外尊贵气质。巷道是一条条归家的小路，摄取自然的韵律，花草景石体现了伊犁四季不同的景色。

NEW CHINESE STYLE
新中式风格

KEY WORDS 关键词

LANDSCAPE MATERIAL
造景材料

THEME DESIGN
主题造型

LANDSCAPE NODE
景观节点

Location: Shenzhen, Guangdong
Developer: Baoneng Real Estate Co., Ltd.
Landscape Design: Shenzhen DongDa Landscape Design CO.LTD
Area: 60,000 m²

项目地点：广东省深圳市
开 发 商：宝能地产股份有限公司
景观设计：深圳市东大景观设计有限公司
项目面积：60 000 m²

Taigu City Garden Residence, Shenzhen
深圳太古城花园

FEATURES 项目亮点

Modern skills and materials combine with traditional Chinese garden style to create a unique park-like environment.

设计以现代的造型手法和材料表达传统中式园林的形式与意境，在造景材料、景观造型以及主题运用等方面特点突出。

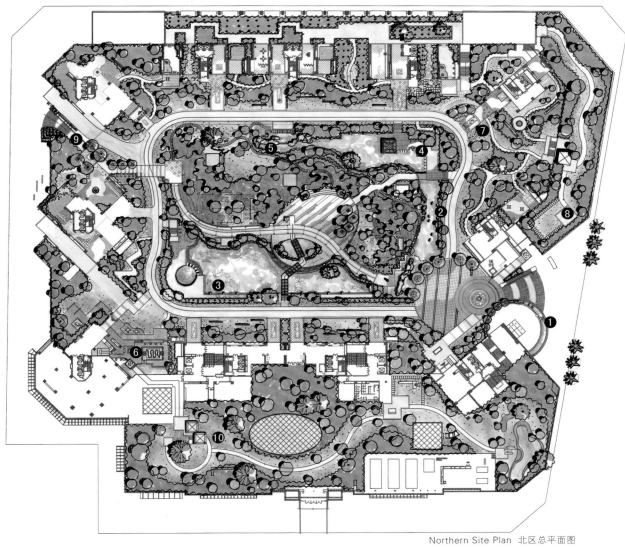

Northern Site Plan 北区总平面图

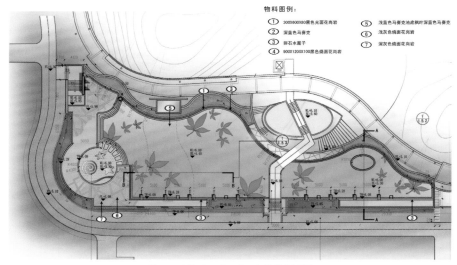

Plane of Swimming Pool 泳池平面图

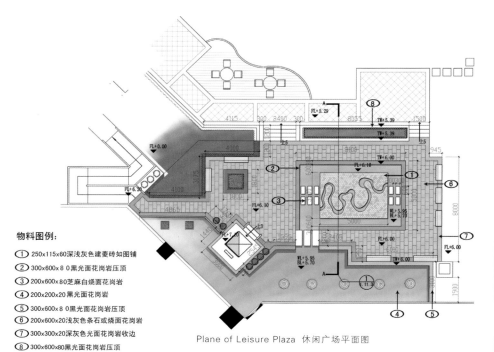

物料图例：
① 250×115×60深浅灰色建菱砖如图铺
② 300×600×80黑光面花岗岩压顶
③ 200×600×80芝麻白烧面花岗岩
④ 200×200×20黑光面花岗岩
⑤ 300×600×80黑光面花岗岩压顶
⑥ 200×600×20浅灰色条石或烧面花岗岩
⑦ 300×300×20深灰色光面花岗岩收边
⑧ 300×600×80黑光面花岗岩压顶

Plane of Leisure Plaza 休闲广场平面图

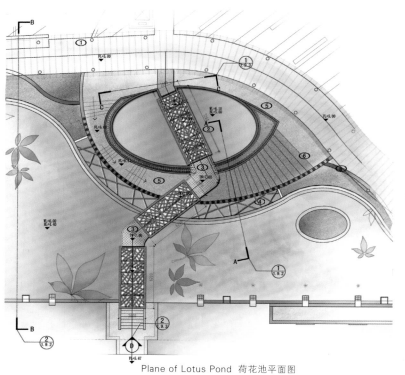

Plane of Lotus Pond 荷花池平面图

Design Concept 设计理念

It is a residential landscape project drawing the quintessence of traditional Chinese garden design with modern modeling methods and materials to highlight the contemporary fashion, which is well integrated into traditional elements and garden architectural forms.

设计汲取了中式古典园林的造园精髓，通过现代的造型手法和材料重新演绎，在继承传统园林内在精、气、神的同时，突出现代潮流及时尚感，努力营造高档、现代、新中式、音乐、自然山水园林式住宅，以现代设计手法及材料表达传统中式园林的形式与意境。

物料图例：
① 特制弧形玻璃顶
② 方通外喷灰色氟碳漆
③ 绿色钢化玻璃
④ 钢板刷黑色氟碳漆
⑤ 300×300×20灰色烧面花岗岩
⑥ 烧面花岗岩上黑色拉丝
⑦ 200×200×20光面黑色花岗岩
⑧ 350×350×80光面黑色花岗岩

物料图例：
① 300X300X20烧面浅灰色花岗岩
② 300X600X20光面黑色花岗岩
③ 300X300X20烧面深灰色花岗岩
④ 300X600X80光面黑色花岗岩
⑤ 卵石水篦子
⑥ 900X1200X100烧面黑色花岗岩
⑦ 300X300X20烧面浅灰色花岗岩
⑧ 兰花状图案，蓝色马赛克
⑨ 250X200X20黑色光面花岗岩
⑩ 渐变蓝色马赛克
⑪ 300X600X20光面黑色花岗岩
⑫ 棕色芬兰木板
⑬ 300X300X80黑色光面花岗岩
⑭ 200X600X80光面黑色花岗岩

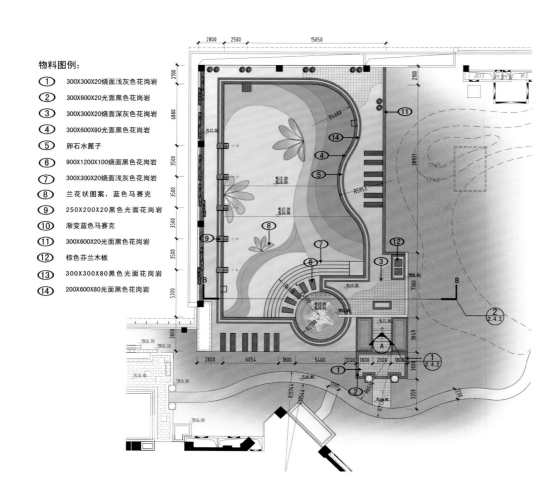

New Materials 新材料的运用

Many modern decorative architectural materials are widely used in this project, such as glass, square steel, stainless steel and granite, to show brand new and special Chinese garden style with traditional garden building forms. A green-glass piled up water-falling rockery is set on the entrance plaza, which gives full express to traditional Chinese rockeryscape by lighting. In the north of this neighborhood unit, a meandering long corridor is covered by a huge piece of arc glass, and the corridor itself is a steelwork. Also, the water-falling gallery and screen decoration on the pool side are built by light steel materials. The decorative wallscape at the empty space is adorned by weaving steel wire. The pavement of this project is made of black-white granite. All pieces of decorative furniture are embellished by stainless steel edges to reflect the minimalism. Modern materials are one of the main elements in this project instead of traditional ones.

深圳太古城花园通过对玻璃、方钢、不锈钢、花岗岩等现代建筑装饰材料在景观构筑物、铺装上的运用，利用传统的中国园林图案和形式的表达，诠释出全新的中式韵味。小区入口处的玻璃跌水假山，以通透的绿玻璃层层叠加，配合隐藏的灯光设备，表达出传统的中式假山景观。北区曲折的长廊，以整块的弧形玻璃定制顶替换了通常的砖瓦顶，并以现代的钢结构取代了传统的木梁结构。泳池雨廊及池边的屏风装饰将惯用的木质雕刻改为轻钢材料。架空层的装饰景墙也通过钢丝的相互编制沿袭出传统竹编制的装饰效果。铺装材料上以黑白灰渐变色的花岗岩代替传统的灰砖、青石。园建小品的不锈钢围边，则用简洁的形态重新勾勒出古典家具的复杂线脚。

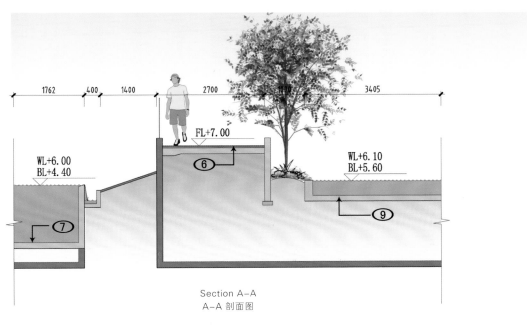

Section A-A
A-A 剖面图

物料图例：

① 特制弧形玻璃顶
② 方通外喷灰色氟碳漆
③ 绿色钢化玻璃
④ 钢板刷黑色氟碳漆
⑤ 300×300×20灰色烧面花岗岩
⑥ 烧面花岗岩上黑色拉丝
⑦ 200×200×20光面黑色花岗岩
⑧ 350×350×80光面黑色花岗岩
⑨ 200×200×20光面黑色花岗岩

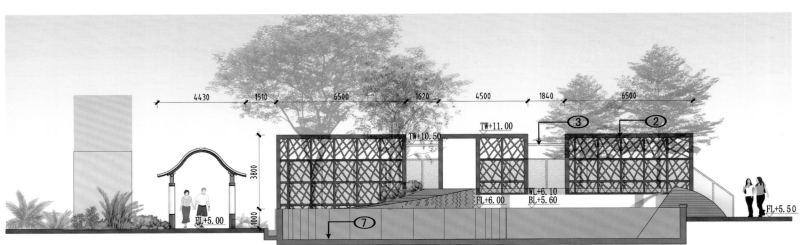

Section B-B
B-B 剖面图

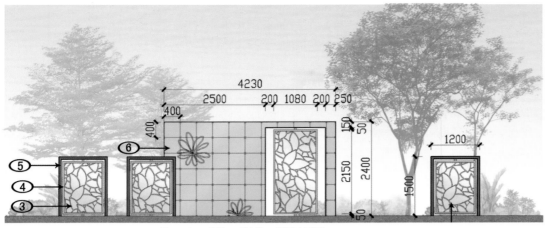
Elevation of View Wall 1 景墙立面图一

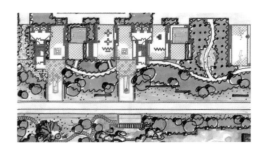

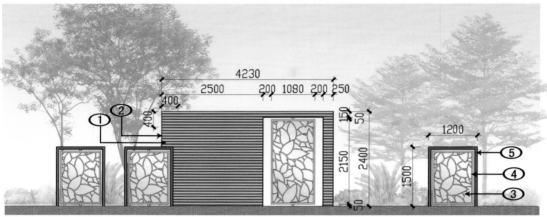
Elevation of View Wall 2 景墙立面图二

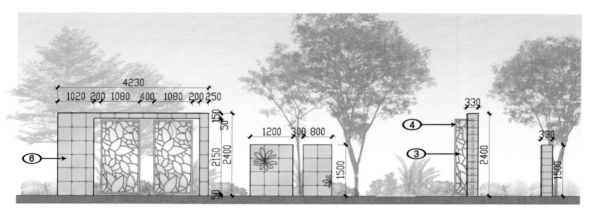
Elevation of View Wall 3 景墙立面图三

物料图例:
1. 50×900光面灰色拉丝花岗岩
2. 20×900光面黑色花岗岩
3. 4厚钢板,刷深灰色氟碳漆
4. 方钢,刷深灰色氟碳漆
5. 方钢,刷黑色氟碳漆
6. 400×400镶嵌式玻璃

Novel Landscape Elements 新颖景观元素造型

The modern design of traditional Chinese garden patterns is another specific in this project. Unique Chinese paper-cut for window decorations on the large-scale ice-crack paned doorframes at the empty space and the allover-patterned pavement both give full express to the integration of modern and tradition.

对中式传统图案的再次设计加工,成为太古城花园的另一亮点。运用在架空层门框上的中式窗花图案,在大尺度的冰裂窗格中利用钢丝进行不同方向填充,形成虚实对比,丰富了原有图案的变化。地面铺装的碎拼图案,则通过演绎变形及对花岗岩毛面拉丝的材料处理,形成不同的图案纹理,演绎出梅花三弄、玉楼春晓、阳春白雪等中式意境。

New Landscape Theme 新主题的运用

Music is the core theme in this project. Each scenery spot here is called by the names of traditional Chinese music, such as The Elegant Music, Orioles Singing in the Willows, The Roaring of Waves in Fall in the north of the site; Blue Waves and Thermal Spring Water, Fishermen Singing the Night Song, Autumn Moon on a Placid Lake in the south part and so on. All these scenery spots are connected one by one as a whole through the unitary element— "water" to form a complete Garden Song. Thus, the project is also a perfect minglement of tradition Chinese music and garden art.

音乐是太古城花园的主题,因此在小区景点的设计命名上,也分别以中国传统乐曲命名。北区高山流水、柳浪闻莺、秋水龙吟、玉楼春晓、阳春白雪、曲水流觞、梅花三弄、平沙落雁和双凤朝阳的九大景点及南区碧涧流泉、渔舟唱晚、平湖秋月、阳关三叠、寒鸦戏水、幽兰逢春、三潭印月、渔樵问答的八处景致,各述其境、缓急不一,并通过"水"这一元素将各景点——串联,形成包含序曲、开幕、过渡、高潮、小高潮、尾音的一幕大型中式园曲。中国乐典韵律之美与中式传统园林之美和谐交融于太古城花园的景观艺术设计之中。

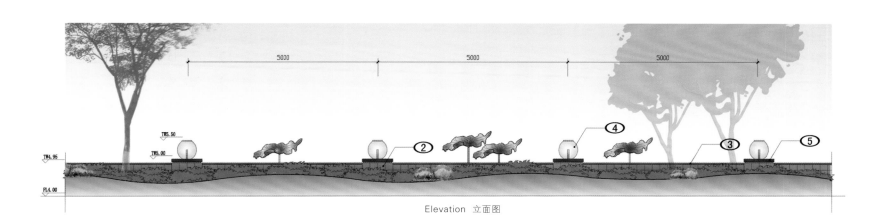

Elevation 立面图

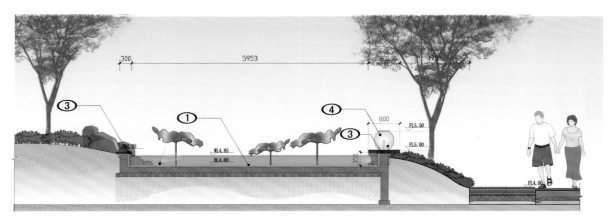

Sectional Drawing 剖面图

物料图例：

① 200X200X20黑色光面花岗岩
② 200X200X20灰色烧面花岗岩
③ 600X300X80光面黑色花岗岩压顶
④ 磨砂玻璃
⑤ 200X200X100黑色光面花岗岩

NEW CHINESE STYLE
新中式风格

KEY WORDS 关键词

ECOLOGICAL TRAITS
生态特质

LANDSCAPE ELEMENTS
景观元素

LANDSCAPE NODE
景观节点

Location: Qingpu, Shanghai
Developer: Shanghai Zhufu Real Estate Development Co., Ltd.
Architectural Design: C.Y. LEE PARTNERS ARCHITECTS/PLANNERS
Landscape Design: Shanghai MEME Landscape Engineering Co., Ltd.
Chief Designer: Zhu Liqing, Li Changyan
Total Floor Area: 126,000 m²
Plot Ratio: 0.17
Greening Ratio: 80%

项目地点：上海市青浦区
开 发 商：上海住富房地产开发有限公司
建筑设计：台湾李祖源建筑师事务所
景观设计：上海唯美景观设计工程有限公司
主创设计：朱黎青 李昌艳
总建筑面积：约 126 000 m²
容 积 率：0.17
绿 化 率：80%

Oasis Lakeside Villa, Shanghai
上海绿洲江南园

FEATURES 项目亮点

Taking advantage of those abundant landscape elements such as river, lake, island, peninsula, bridge and other elements, a landscape full of Jiangnan ancient melodies and poetry has been created.

利用河、湖、岛、半岛、桥等丰富的景观元素，塑造出充满江南古韵和诗情画意的景观。

▶ Overview 项目概况

The project is adjacent to the living community of Zhujiajiao and the 533,333 m² Dading Lake. Lakeside villa, where the project locates, covers an area of 266,667 m². Each household varies from 295 m² to 645 m² and the private garden varies from 1,000 m² to 2,000 m², there are 20 kinds of housing type, and 300 households.

绿洲江南园紧临上海青浦区朱家角古镇生活区，临近 533 336 m² 的大淀湖。这块始建于 2002 年的湖景别墅区，总面积约 266 670 m²，每户面积在 295 m²~645 m² 之间，私家庭园面积 1000 m²~2000 m²，共有房型 20 余种，总栋数 300 户。

▶ Design Concept 设计理念

The project is located at the oldest Shanghai millennium town, Zhujiajiao (one of the four historic towns), a tourist spot for heads of APEC, the meeting place for Shanghai International Tourism Festival, the cradle land of Songze Culture and the preservation site of world cultural heritage. This historical cultural background aroused designers' pursuit for residential culture of southern Chinese riverside town. In addition to the enchanting scenery, charm of waterfront living atmosphere is introduced, which brings about a brand new feeling about cultural life for the residents. Bluish-grey shed roof and the stone facade set a new banner in Shanghai villa.

绿洲江南园地处上海历史最久远的千年古镇朱家角（四大历史名镇之一），这里孕育的别墅居住文化，是典型的江南水乡居住文化。绿洲江南园在尽享湖光水色的同时，还融入了浓郁的水乡生活氛围，使居住者对文化生活产生新的感觉。建筑风格为现代中式风格，这样与古镇风貌建筑群协调。蓝灰色的单坡屋顶及全石材立面，在上海别墅区可谓别树一帜并独显尊贵。

Characteristic Landscape 景观特色

In the respect of landscape design, this villa area boasts two features. One is the profound unmatched cultural atmosphere, and then the 555,333 m² Dading Lake and the surrounding ecological park. The villa area is the so-called "yacht villa", for it is comprised of more than 20 peninsulas, which provides thoroughfare for small private yachts.

In the respect of planning layout, it makes the best of landscape elements such as river, lake, islands, peninsulas and bridge to construct a picturesque residential area with Jiangnan ancient charm. Nowadays, European style is incredibly well known. Under the circumstances, this ecological and antique landscape is just like an adorable fresh wind.

该别墅区景观规划设计的重要特点，一是深厚的江南古镇文化氛围，这点无以复制；二是包围着大淀湖及大淀湖生态园，别墅区大约由二十处半岛组成，户户枕水，可通行小型私家游艇，也就是所谓的"游艇别墅"。

设计规划利用河、湖、岛、半岛、桥等丰富的景观元素，塑造出一处地形蜿蜒起伏、河水荡漾妩媚、植树浓密匝地、景桥姿态万千、垂柳婀娜多姿、竹子亭亭净直的充满江南古韵和诗情画意的景观。在欧陆风盛行的年代，这种清新的江南水韵景观，似乎是一股清新之风，分外妖娆，惹人怜爱。

NEW CHINESE STYLE
新中式风格

KEY WORDS 关键词

ECOLOGICAL SPECIALTIES
生态特质

IDEAL CONDITION CREATION
意境营造

LANDSCAPE NODE
景观节点

Location: Guangzhou, Guangdong
Developer: Guangdong Zhongli Investment Company
Landscape Design: Guangzhou Sansui Bidder Landscape Design Co., Ltd.
Planning Area: 184, 000 m²
Landscape Area: 118, 000 m²

项目地点：广东省广州市
开 发 商：广东中力投资有限公司
景观设计：广州山水比德景观设计有限公司
规划面积：184 000 m²
景观面积：118 000 m²

Guangzhou Dayi Villa
广州大一山庄

FEATURES 项目亮点

The design principle is put forward through the double level of sense and aesthetics, strive to let ideal condition, culture, art and nature be in harmony and unity.

项目从理性与美学双重层面提出设计原则，力求在意境、文化、艺术上与自然和谐统一。

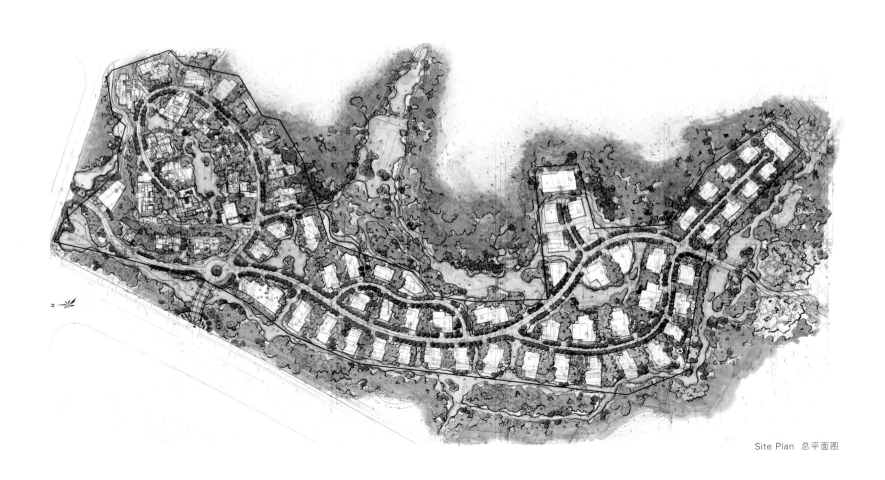

Site Plan 总平面图

Overview 项目概况

The Project named "Dayi" means to pursue a state that harmony between man and nature, unification of natural law— Add Yi to Da means Tian, subtract Yi of Da means Ren. Hence, the positioning of this project is that landscape should only be in heaven, falling into the world is fairyland—the concept of Poetry world.

项目名曰"大一",意在追求天人合一、天道归一之境界——大之加一谓之"天",大之减一谓之"人"。因此,在项目的定位上,便有"此景只应天上有,落入人间为仙境——诗意人间"的立意。

Design Concept 设计理念

The design proposal combined orientation of this project with client's demand and site resource. The design concept originated from Peach-Blossom Spring, sublimates artistic conception, culture and spirit, and formed the idea of this project planning which is Moon Water Cloud, the Peach Garden—Heaven on Earth.

结合项目定位、客户需求以及场地资源等情况,设计方案也是呼之欲出。经过思想的碰撞、心灵的感悟,承袭项目之初"水、月、层、云"的理念和"天法道,道法自然"的原则,感悟"水、月、层、云、林、艺、奇、然"的景观意向,本方案以《桃花源记》为引子,在意境、文化、精神上进行完美的升华,形成了本次项目规划的理念:水月云天·桃花源——天上人间。

Design Principle 设计原则

The designers investigated the vertical and horizontal plane elaborately in the process of general layout of landscape: a waterfall was planned among mountains through uniting the river and mountain of the site, which formed the momentum that the Silver River fell down from azure sky. The waterfall flowed through the whole community and the drainage system in the community is jade belt, to enhance the interaction within community via "Jade belt legend". The central water in this housing cluster developed into a pattern of "Seven-stars with Moon", made the whole water system developed into an organic entity.

在景观水体的总体布局中,设计师在竖向和平面上都进行了详尽的考究:结合项目基址的山形水势,在群山之间规划一条瀑布,形成银河落九天之势;瀑布之水流向整个社区,形成的水系即为玉带,用"玉带船说"的形式来加强社区内部的互动交流;组团之间的集中水面形成"七星拱月"之势,使水体系统形成一个有机的整体。

▶ Landscape Layout 景观布局

Moon stands for a kind of realm, or complex in China! The design of this project brought in the concept and complex of moon for the sake of interiorizing the concept of home, to make Dayi Mountain Villa be a home that the owners long for most. The features of clouds are thousands of shape, boundless imagination. The designers built an atmosphere of cloud through dealing with fog spring in designing the landscape. Tian means harmony between man and nature, heaven on earth in this project. Dayi Mountain Villa contained a realm that natural law united, the connotation of "Dayi" is all-inclusive, with the metaphor of a state that harmony between man and nature is the heaven on earth.

The design principle of the project pointed out form double level of sense aesthetics: the perspective principle is based on current situation, high starting point and high standard, strived to unification of nature and art, integration of environment and humanity. Ecological sustainable development principle should consider the balance between present and future, manage the relationship between natural resource protection and development and construction, plan scientifically, distribute rationally; the principle of peculiarity and humanity united, to let this garden be filled with poetry by those mode of artistry, literacy, spiritualization and nature ecologicalization, transmit infinite lyric imagination from finite buildings; The principle of people-oriented emphasized on participation of people from different level, different age grades, the affinity, recreation and entertainment with water.

　　月，代表的是一种境界，一个情结。本案设计中将月的概念和意境引入到景观之中，就是要让家的概念深入人心，让"大一山庄"成为业主心里最盼望回归的家；云，层云叠影，形色万千，无穷想像，景观设计中通过雾泉的处理方式为社区的整个意境营造这种氛围；天，意即天人合一、天上人间。"大一山庄"蕴含着天道归一之境界，"大一"之中更是包容万象，暗喻天人合一，是人间的仙境。

　　整个项目的设计原则，从理性与美学双重层面提出。前瞻性原则：基于现状，高起点、高标准，力求自然与艺术的统一、环境与人文的统一；生态可持续发展原则：考虑协调当前与未来的平衡，正确处理自然资源保护与开发建设的关系，科学规划、合理布局；独特性以及人文性相结合的原则：以艺术化、文学化、心灵化、自然生态化，使园林渗透、充盈着诗意或者文心，从有限的建筑传达出抒情的无限想像；以人为本原则：意即强调不同层次、不同年龄阶段人群的参与性、与水的亲和性、休闲性与娱乐性。

NEW CHINESE STYLE
新中式风格

KEY WORDS 关键词

INK ELEMENTS
水墨元素

COURTYARD LANDSCAPE
庭院景观

LANDSCAPE NODE
景观节点

Location: Yangjiang, Guangdong
Landscape Design: Guangzhou Yuanmei Environmental Art Design Co.,Ltd
Designer: Wu Yingzhong

项目地点：广东省阳江市
景观设计：广州市圆美环境艺术设计有限公司
设 计 师：吴应忠

Yangjiang Danmo Cabin
阳江淡墨幽居

FEATURES 项目亮点

Using ink as its main elements, with "the beautiful scenery of mountain, river, and island is like a painting, green leaf, red flower and grass as if have emotion" as landscape artistic conception.

采用水墨为主元素，营造"远山近水群岛若画里，碧叶朱花小草皆有情"的景观意境。

Overview 项目概况

This project is located in northern Xinhua Road, eastern Longtan Road in Yangdong county, Yangjiang City. The design is the landscape part of Danmo Cabin which belongs to Huiqinlin Villa life-preserving garden, and covers an area of 2,200 m², and garden landscape design area of 600 m². Huiqinlin Villa covers an area of 180,000 m², with a total floor area of 300,000 m², The villa positioning is "leading the noble life in western Guangdong", and the project positioning is "leading luxury mansion life in Yangjiang", building a low density, high quality and high grade complex villa community in Yangjiang, The type includes garden villa, townhouse and eight high-rises.

  本项目位于阳江市阳东县城新华路以北、龙塘路以东，此次设计为中惠沁林别墅豪宅养生庭院——"淡墨幽居"的景观设计部分，占地面积 2 200 m²，庭院景观设计面积 1 600 m²。项目所在的中惠沁林山庄总占地面积 180 000 m²，总建筑面积 300 000 m²，产品定位是"引领粤西名门生活"，项目定位是"引领阳江豪宅生活"，在阳江市打造一个低密度、高品质、高档次的复合型社区，类型有花庭美墅、连体别墅和八栋高层。

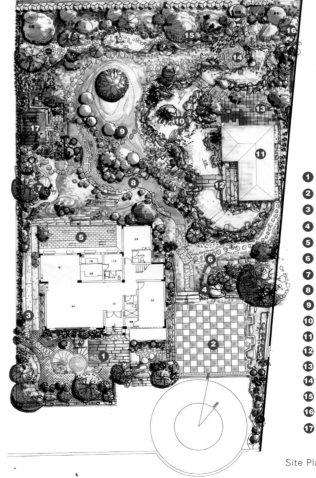

1. 客似云来
2. 六合阴晴
3. 红叶寄情
4. 淡墨幽香
5. 万籁无声
6. 归来悟空
7. 笑傲风月
8. 蛟龙得水
9. 风轻云淡
10. 沧浪之水
11. 墨趣轩
12. 双源桥
13. 抚琴台
14. 竹林风韵
15. 观山揽翠
16. 飞珠溅玉
17. 归隐田园

Site Plan 总平面图

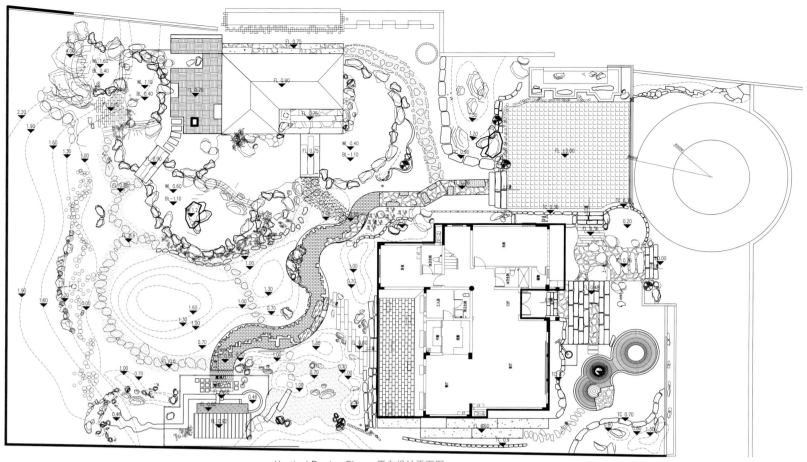

Vertical Design Plane 竖向设计平面图

Design Theme 设计主题

With "the beautiful scenery of mountain, river, and island is like a painting, green leaf, red flower and grass as if have emotion" as landscape artistic conception, ink is used as its main elements. The ink is divided into five colors: "dry, wet, strong, weak, coke". With a Chinese writing brush, a piece of Duan ink, life phenomena, ups and downs, are showed on the paper. Ink presents trace, which is connect to density, weight, the primary and secondary scenery of and landscape design, Water originates from the famous quotation "as good as water, water is good at everything while indisputable". The highest level of good deeds is like the character of water, spread all-round benefit to all but never ask for fame and wealth, water stands for virtue, the experience of heart, invisible but supreme, it seems to walk around an idle court, sit at the corner of the courtyard, taste a cup of tea, meditate for a while.

以"远山近水群岛若画里,碧叶朱花小草皆有情"为构景意境,采用了水墨为主元素,墨分五色"干、湿、浓、淡、焦",一管狼毫,一块端砚,人事百态,人生起伏,尽现纸上。墨代表行为着迹,与景观设计中的疏密、轻重、主次布景相联系;水引自"上善若水,水善利万物而不争"之语,所谓最高境界的善行就像水的品性一样,泽被万物而不争名利,水代表德,心的体验,无形而至高,恰似信步闲庭间,静坐庭院一隅,品一缕茶香,静思沉酿。

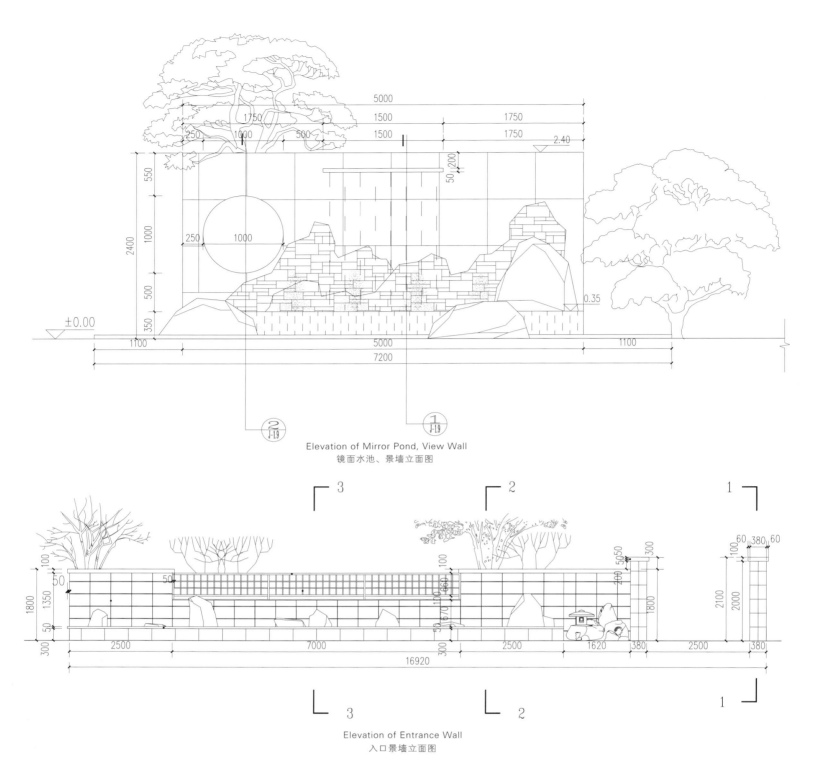

Elevation of Mirror Pond, View Wall
镜面水池、景墙立面图

Elevation of Entrance Wall
入口景墙立面图

Courtyard Landscape 庭院景观

The designer hopes to give some spiritual things to the landscape design, and make the landscape full of vitality, meanwhile, the spirit of landscape should be close to clients' value inorder to achieve such a result when the owners living here, they can feel the yard has a conversation with themselves, draw power from nature, release themselves, open and enlighten their thinking.

整个庭院设计希望将精神层面的内容融入景点中去，让景观具有生命力，同时这种精神是最大化地接近业主的价值观，以使业主在起居之时，能感受到庭院与自己的对话，共呼吸，汲取自然的力量，释放自我，启迪思想。

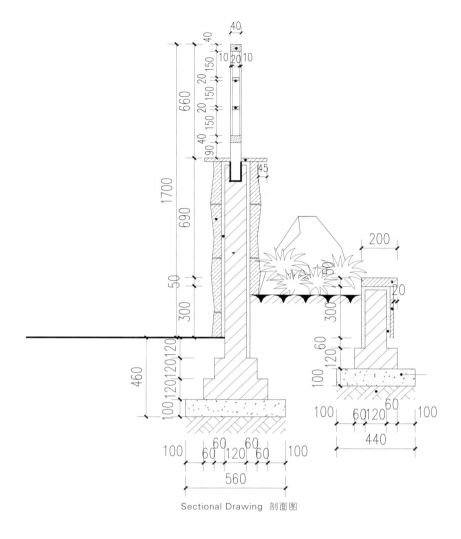

Sectional Drawing 剖面图

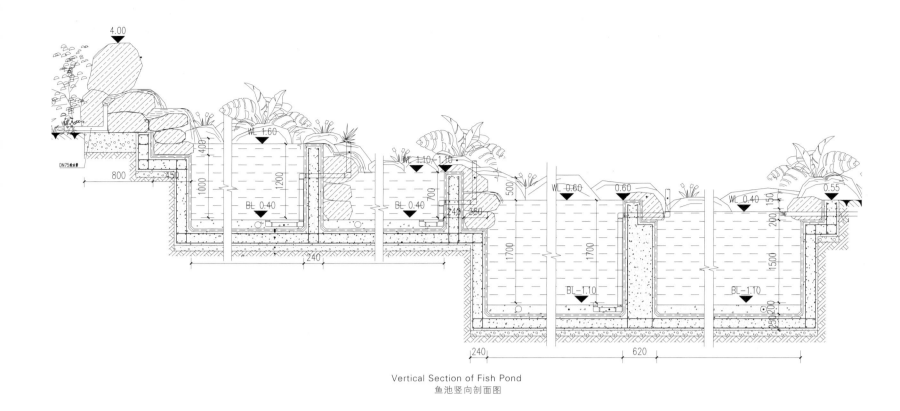

Vertical Section of Fish Pond
鱼池竖向剖面图

NEW CHINESE STYLE
新中式风格

KEY WORDS 关键词

ECOLOGICAL SPECIALTIES
生态特性

SPACE CONTEXT
空间文脉

LANDSCAPE NODE
景观节点

Location: Nanchang, Jiangxi
Developer: Jiangxi Vanke Yida Real Estate Development Co., Ltd.
Landscape Design: Shenzhen ALT Architectural Landscape Design Co., Ltd.
Land Area: 130,000 m²

项目地点：江西省南昌市
开 发 商：江西省万科益达房地产发展有限公司
景观设计：深圳市雅蓝图景观工程设计有限公司
占地面积：130 000 m²

Vanke Rain Garden, Nanchang
南昌万科润园

FEATURES 项目亮点

Preserve the existing landscape resource in the site, combine with architectural planning, integrate with local cultural connotation, achieve a state of sights within sights, living within culture.

项目尊重场地内的现有风景资源，配合建筑规划，并融合当地文化内涵，实现在风景之中生长风景，在文化之畔坐落生活。

▶ Overview 项目概况

Located in Qingyunpu traditional commercial district, former of the site is for public security school. Trees that exist for several years set a great start for the landscape design of this project which cooperates with architectural planning and makes the best of the scenery resource.

  本项目位于江西省南昌市青云谱传统商业街区一侧,原为公安学校用地,用地内有若干多年生长的大树,林荫婆娑,枝桠虬然,宛若水墨意境。项目的景观设计以此为基点,配合建筑规划,并融合当地文化内涵实现在风景之中生长风景,在文化之畔坐落生活。

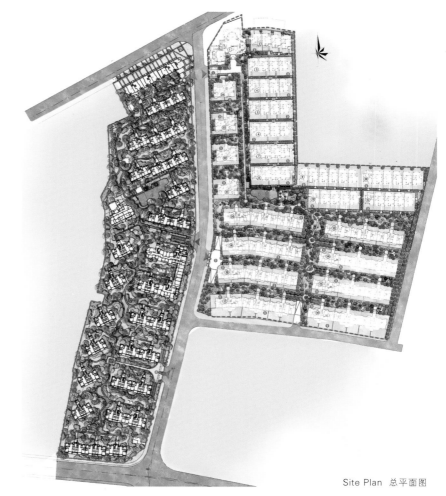

Site Plan 总平面图

Elevation of Community Secondary Entrance
小区次入口立面图

Design Principle 设计原则

In terms of landscape design, it inherits regional context characteristics, connecting different functional and landscape spaces through layering courtyard space, converting atmosphere and mental feelings from urban public space to private space in progressive levels, thus to create fine quality home landscape. The proper functional layout not only meets the need of outdoor activities but ensures the privacy. Vertical height difference has been processed effectively that produces a novel landscape effect. It considers the relation of the landscapes in both east and south areas, connecting integrated scenery with divided community.

景观设计方面，项目传承空间文脉特征，通过层进的庭院空间围合连接不同的功能与景观空间，在递进的层次中转换从城市公共空间到私家专属空间的气质氛围与心理感受，由此营造出具有品质感受的家园景观。尊重场地内的现有风景资源，保留原生大树，并作为社区景观的主要元素。合理布局社区功能空间，既满足户外生活的功能要求，又保证居住的安静与私密。有效处理场地竖向高差，并由此产生新颖的景观效果。兼顾东、西区景观的呼应与联系，使整合的风景联络被分隔的社区。

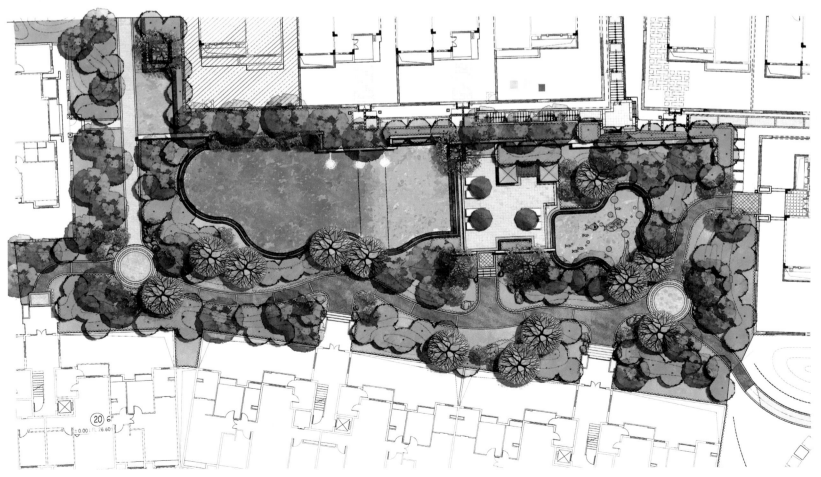

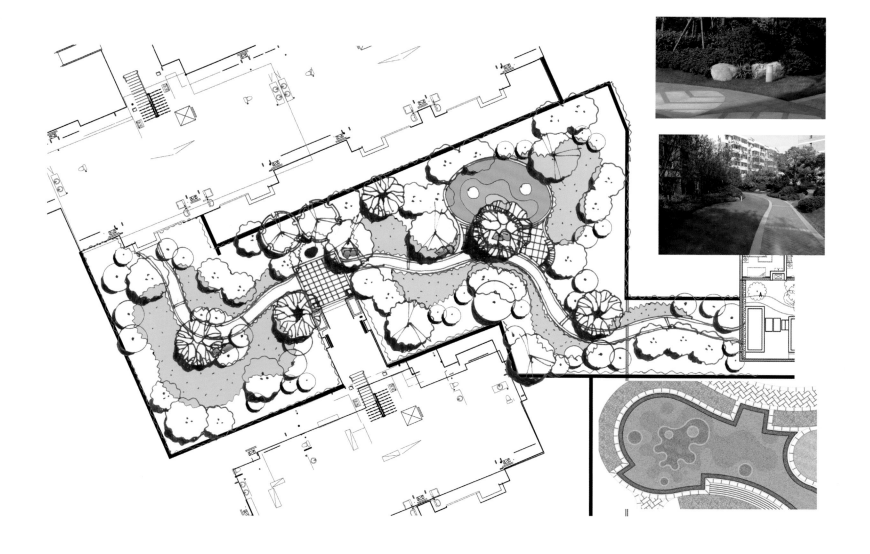
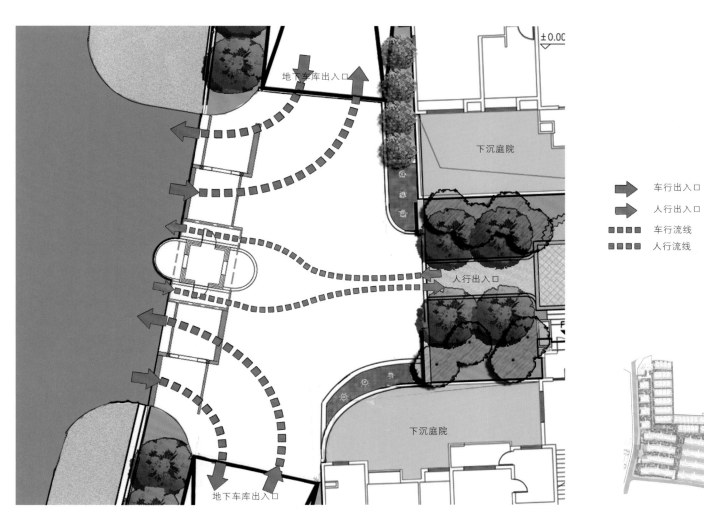

Traffic Drawing 交通分析图

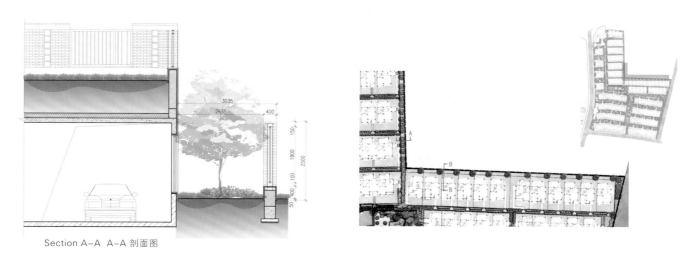

Section A-A A-A 剖面图

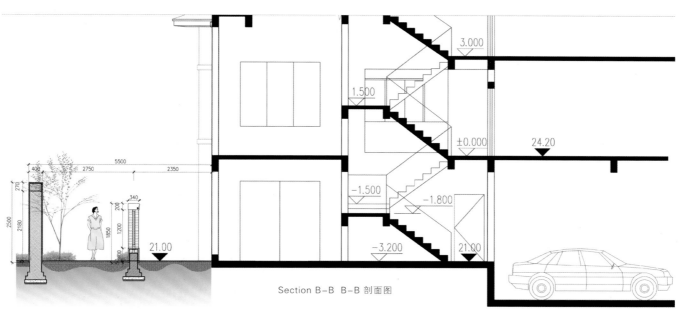

Section B-B B-B 剖面图

Vegetation Design 景观植物配置

In terms of plant landscape configuration, it fully integrates with the characteristics of the north-south narrow site, respecting existing scenery landscape and retaining original trees to green living environment in a true sense. It selects local green plants in according to the conditions of the site and the trees, creating three-dimensional levels by integrating trees, brushes and flowers to express the natural flavor and inherit regional contexts. Natural planting form and profound environment experience came into being in the unified rhythm and limited space. In addition, it pays attention to the seasonal variable characteristics of the plants, in which you can see the fresh green plants all the year around.

项目在植物景观配置上，充分结合狭长的南北走向场地特点，尊重场地内的现有风景资源，保留原生大树，真实地塑造出绿意葱茏的绿色居住环境。以适地适树为原则选用本土特色植物，结合现状大树及观赏大灌木、丛生地被灌木与花卉进行立体层次营造，自然中蕴含野趣，清新处不失质朴，现代而延续文脉。林木种植以自然的树丛结合为主，在统一的节奏中变化出自然的种植形态，在狭窄有限的空间中营造出丰厚的植被感受及幽深的环境体会，设计中同时重视植物景观四季季相的变化特征，景随季变的同时形成丰富的植物空间群落，给人以回归自然的清新感受。

NEW CHINESE STYLE
新中式风格

KEY WORDS 关键词

THEME PARK
主题公园

GREEN SLOPE
坡度绿化

LANDSCAPE NODE
景观节点

Location: Xi'an, Shaanxi
Landscape Design: IDU Architectural Planning and Landscape Design Co.,Ltd.
Landscape Area: 30, 000 m²

项目地点：陕西省西安市
景观设计：深圳IDU（埃迪优）建筑规划与景观设计有限公司
景观面积：30 000 m²

Southern Orchid Garden, Xi'an
西安兰亭坊南区

FEATURES 项目亮点

The landscape lays stress on nature and ecology, using a series of landscape architectures and green slopes to create an eco and natural environment.

设计上强调生态与自然的结合，运用了一系列特色的景观小品及坡度绿化来协调整个空间的自然景观氛围。

Overview 项目概况

The project is located over the same altitude with Tangmuta Temple, the collection site of Lanting Xu, the art treasure of ancient China, and adjacent to the ancient ruin of Chang'an Town of Sui and Tang Dynasties. Assembling the royal momentum and the noble area, the project is built for the modern elites with five ecological parks around.

该项目地处艺术瑰宝《兰亭序》珍藏之地西安唐木塔寺文脉之上，毗邻隋唐长安城坊遗址，集享皇都龙气王脉，承传大唐盛世繁华，坐拥高新稀贵领域，五大公园生态环伺，东方园林的皇家韵律营造当今名士居住的生活境界。

Site Plan 总平面图

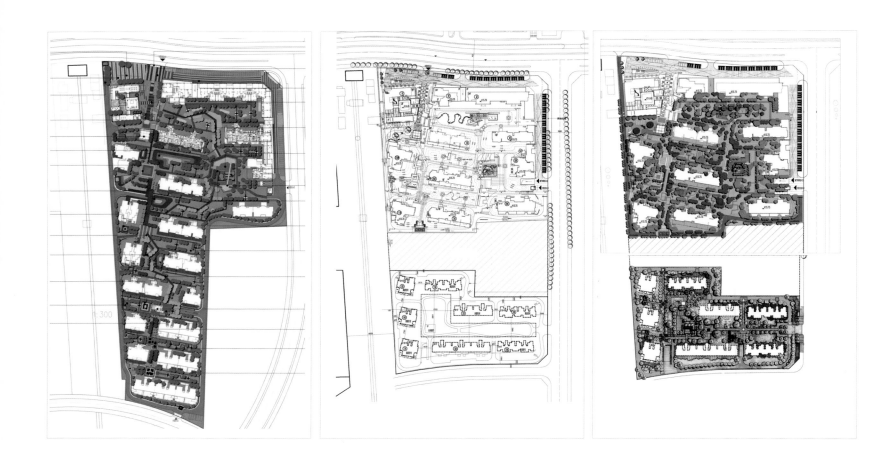

Landscape Feature 设计特色

On the design for the garden landscape, the project lays stress on nature and ecology. The slope greening, wooden stairs, featured potted plants and waterside greening in Dragon Hiding Pavilion have reflected the integration of nature and human. The delight design has brought about the community concept of human and nature respecting in harmony. The five landscape principles are highly appreciated for expressing the true attitude of life. The harmony between human and nature, the coexistence of construction and nature, the friendly relationship among neighborhood are formed to make contribution to the harmonious society.

在园林景观的设计规划中，兰亭坊在园林规划上更注重"自然"和"生态"特色，龙隐阁庭院中的坡度绿化、木楼梯、特色盆栽、龙隐阁、水边绿化、坡度绿化体现了大自然与人的自然融合。设计内外通透，富有情趣，体现出尊重人与自然的"和谐"的社区理念。其五条景观核心价值值得称道，即自然本真的生活态度、人与自然的和谐相处、建筑与自然的和谐共生、和谐融洽的邻里关系、营造的和谐文化是和谐社会的基石。

Landscape Node 景观节点

The three royal theme parks in the Orchid Garden are based on the three festivals of Tang Dynasty. Originated from the cultural background of Spring Dragon Festival, dragon has been respected as the god that brings harvest and good health to people. Therefore, the sights such as Dragon Wall, Dragon Hiding Pavilion, Dragon's Paw Elm and Taihu Lake Stone are set in the theme parks. According to the festival on March 3th in the lunar calendar, the sights of Emotion in Landscape, Personage Manner and Winding Stream are built. While the sights of Forever Pavilion, Double Ninth Tree Array and Chrysanthemum come from the Double Ninth Festival.

The southern part of Orchid Garden is newly built by four sections: ruin protection, leisure commerce, culture and arts and central lake. A comprehensive park of arts, leisure and nature is formed. The northern entrance follows the spool thread of south-north direction and combines with the cultural arts section on the west and the leisure commerce section on the east, forming the natural landscape section subjecting culture of ruins.

The center is set with a landscape lake surrounding by a wooden bridge, in the center of which there is a music fountain, letting the residents feel the cultural and artistic atmosphere of southern area when walking around the lake. As a wide open entrance square, the southern entrance is set with scattered tall landscape columns which present historical and modern integration sense in the light and correspond with the Tang ruin park on the south side.

Additionally, Orchid Garden is located in the high-tech new district with diversified living facilities. The creative industry area, conference economy, dozens of hotels and the administration center have formed the rich zone of Xi'an.

兰亭坊社区内的三大皇家节日主题公园以营造盛唐三大节日作为契机。中和节主题公园,以"二月二,龙抬头"作为文化背景,意为引龙,一是请龙,祈求农业丰收;二是龙作为百虫之神,对人体健康、农作物生长都是有益的。当中有龙壁、龙隐阁、龙爪榆、太湖石等景点。上巳节主题公园中的"上巳"最早出现在汉初的文献,日子为"三月三",《周礼》郑玄注:"岁时祓除,如今三月上巳如水上之类。"主题公园创造了寄情山水、名士风范、曲水流觞等景点。重阳节主题公园以农历"九月九"的重阳节打造了久久亭、重阳木树阵、赏菊等景点。

兰亭坊南部新建区由遗址保护、休闲商业、文化艺术及中心景湖四个部分构成,形成一个集艺术、休闲、自然为一体的综合性公园,北入口继承南北轴线关系,与西部文化艺术区、东部休闲商业区相结合,形成以遗址文化为主的自然风景区。

其中,中心部分设木桥环绕穿越的景观湖,在湖中央,有一组音乐喷泉,使人们绕湖散步之余能体验到南区的文化艺术氛围。南入口为一个开敞的入口广场,分散布置着高大的景观石柱,在灯光的映射下展现出历史和现代的交融感,并与南边的唐遗址公园形成呼应。

此外,兰亭坊身处高新区,多种多样的高尚的生活配套次第展开,创意产业带、会展经济、数十家酒店、高新行政中心,形成了西安的智富区域。